Engineering, Business and Professional Ethics

Simon Robinson, Ross Dixon,
Christopher Preece and Krisen Moodley

Routledge
Taylor & Francis Group
LONDON AND NEW YORK

First published by Butterworth-Heinemann

First published 2007

This edition published 2011 by Routledge
2 Park Square, Milton Park, Abingdon, Oxfordshire OX14 4RN
711 Third Avenue, New York, NY 10017

Routledge is an imprint of the Taylor and Francis Group, an informa business

First issued in hardback 2015

British Library Cataloguing in Publication Data
A catalogue record for this book is available from the British Library

Library of Congress Cataloging-in-Publication Data
A catalog record for this book is available from the Library of Congress

ISBN 978-0-7506-6741-8 (pbk)
ISBN 978-1-138-13814-8 (hbk)

Contents

Introduction

The skills of ethical reflection are central to the practice of engineering and management. Sometimes these reflections are practised via legislation, such as in the area of Health and Safety, or the process of consultation as a major project is developed. Sometimes ethics simply emerge in the context of a project because of work-based practices that are questionable, because of conflicts that arise out of basic ideas, purpose or values. The case of the Challenger Flight 51-L (referred to as the Challenger, which will be examined in some detail in this book) is a good example. In that, engineers and managers looked at problems from very different perspectives, and the influence of these, far from enabling the contrasting and conflicting values to be explored and worked through, led to disaster.

Sometimes ethical issues may arise out of a clash of cultures. A firm's policy of not accepting gifts may come up against accepted practice in a country where this is seen as normal. Indeed, it may be seen as unethical to refuse a gift. Reflection involves finding the best ways of recognizing and dealing with such issues, and this process might require a closer look at the values of the individual, organizations or culture in question and how such differing values can be appreciated and most creatively accommodated.

Such issues are faced by engineers and managers on a regular basis, and, with the growth and spread of multinational corporations, they are faced with ever more complicated and challenging situations and dilemmas. In practice they often get on with the response to a situation without actually identifying or even recognizing any aspect of it as 'ethics'. The result has been that the engineering professions have not, unlike Medicine or the Law, developed a strong body of history or training on Applied Ethics. Medicine and Law in particular have seen the creation of almost a separate, high-profile discipline of applied ethics. This is probably because their work frequently involves headline issues, such as euthanasia, abortion or the fate of conjoined twins (Lee 2003). The latter case is a fascinating example of very different issues, such as right to life, the definition of the individual, responsibility for making decisions and the several different perspectives of those involved, from the Roman Catholic Church, to the judges, to the parents, to the twins themselves.

Engineering and Management Ethics do not at first sight have the same profile, partly because they are not apparently faced with the same complexity of dilemma. Yet engineering does raise key issues, sometimes with much greater consequences, affecting large groups of people, sometimes even society and the environment. An example is the Chernobyl disaster, which has quite rightly been set down in the history of engineering as an event to be studied and from which valuable lessons must be learned. A review of such projects shows how any technical decision is set in a context of values, and how values and attitudes can affect the way in which engineering professionals actually see their situation.

Alongside the high-profile projects and disasters are the normal everyday issues. These too can hold ethical challenges, part of the complex demands of the profession. We hope to show that much can be learned from both types of experiences in this book.

This book intends to offer the reader as a professional engineer, manager or engineering manager how to:

- Develop a practical method of ethical decision-making.
- Think holistically and proactively so that ethical reflection becomes part of professional practice, and not simply a bolt-on extra to professional practice.
- Become familiar with ethical theory and different belief systems that inform engineering and management practice.
- To explore and understand the particular issues in engineering management ethics.

Philosophers are often wary of method-based approaches. Just to have a method, it is argued, discourages taking responsibility for the ethical reasoning, and means there is little discussion of what ethics actually means – what is right and wrong and they might be justified. Practitioners, on the other hand, are wary of the academic philosopher or theologian. They are perceived to have many wonderful theories about right and wrong but their heads are simply not in the 'real world' of the engineering project. There is often a fear on both sides that they are being judged by the other.

We hope to establish a shared discourse that values both theory and practice. Professional ethics textbooks all too often do not take seriously the relationship between ethical theory and practice. They tend to begin with theories, and then move into reflection on practice that does not embody the meaning of that theory. We aim to be practice centred, with theory informing that practice.

The book is intended to be a journey of discovery through ethics, professionalism, engineering and management. Chapter 1 looks at the reasons for 'being ethical'. It notes that there are many good self-interest-based reasons for being ethical, but that these are not sufficient in themselves. The second part of this chapter explores the sense in which being ethical is part of what it means to be a professional engineer, and how this differentiates the engineer from a technician. The more we ask the question 'why should I be ethical', however, the more we find that we need to know just what this 'ethics' comprises. Hence, Chapter 2 explores the meaning of ethics, the underlying theories of ethics – noting their importance and also their limitations. We look at ethics as part of the broader reflection on practice that seeks to make meaning, and the different ways in which meaning is made. This moves us from regarding ethics seen simply as solving puzzles about right and wrong to ethics as concerned about character, with words like integrity brought to the fore. From this understanding we will look at the core virtues, those capacities that enable all of us to be aware of ethical issues and respond to them. We suggest what the virtues of the professional engineer might be and end the chapter by offering some ethical operational principles.

Chapter 3 then looks at ethical methodology, asking 'how do you do ethics'. If ethics is not simply to be about the unthinking application of rules or principles then we need capacity to respond appropriately to any challenge and also a method that will enable this response. Towards these ends we begin by focusing on reflective practice, then developing an ethical methodology. We will then show the ethical dimension of the engineering project.

Chapter 4 investigates the professional codes comparing several standard ethical codes, reflecting on what codes are for, what they should contain and how they should be used. The relationship of codes to practice and the identity of the professional community are then developed.

The second part of the book moves from engineering and professional ethics to the context of the ethics of the engineer in business. Chapter 5 looks at the basis of business ethics, noting the importance of taking it beyond simply individual decisions in business. Chapter 6 focuses on the details of corporate social responsibility. Chapter 7 develops environmental ethics and sustainability. Chapter 8 takes the business into the area of global ethics. Increasingly, the issues of environmental and global ethics are forcing businesses, small and large, into becoming aware of the social and physical environment and of the need for their practice to be transparent.

Chapter 9 draws these issues together and sets them in a consideration of the wider philosophy of applied science. We hope that the book as a whole will set the values of engineering into a meaningful context, and above one that will make a difference to practice.

Finally, we would like to thank three groups of people without whom this book would not have been possible: our students who constantly challenge us on ethical issues, our colleagues in academia and industry who have constantly asked the awkward questions and our loved ones who have not complained about long absences from the domestic scene and, indeed, strangely seem to have quite enjoyed them.

July 2006

REFERENCE

Lee, S. (2003). *Uneasy Ethics*. London: Pimlico.

Authors

Rev. Prof. Simon Robinson, MA, BA, PhD, FRSA

Professor of Applied and Professional Ethics, Leeds Metropolitan University, Associate Director, Ethics Centre of Excellence, and Visiting Fellow in Theology, University of Leeds.

Educated at Oxford and Edinburgh universities, Simon Robinson entered psychiatric social work before being ordained into the Church of England priesthood in 1978. After spells in the Durham diocese, he entered university chaplaincy at Heriot-Watt University and the University of Leeds, developing research and teaching in areas of applied ethics. In 2004 he joined Leeds Metropolitan University. Ongoing research interests include professional ethics; ethics in higher education; spirituality and professional practice; corporate social responsibility and ethics in a global perspective. Books include *The Social Responsibility of Business; Ethics in Engineering; Agape, Moral Meaning and Pastoral Counselling; Case Studies in Business Ethics* (edited with Chris Megone); *Living Wills; Spirituality and Healthcare* (with Kevin Kendrick and Alan Brown); *Ministry amongst Students; Values in Higher Education* (edited with Clement Katulushi); *The Teaching and Practice of Professional Ethics* (edited with John Strain); *Ethics and Employability; Spirituality and Sport.*

J. Ross Dixon, BSc, CEng, MICE

Ross Dixon is a Chartered Civil Engineer with over 30 years' experience in the construction industry. After training with a major civil engineering contractor he has worked in consulting engineering companies as Resident Engineer, Design Engineer, Chief Civil Engineer and Technical Director in practices based in the United Kingdom and Hong Kong, Bahrain, Qatar, Saudi Arabia, UAE and the Philippines.

He has broad experience in the preparation of feasibility studies and the planning, design, construction and project management of building and civil engineering works.

He retired in 2005 from the post of Senior Lecturer in the Construction Management Group of the School of Civil Engineering, University of Leeds, where he currently lectures part-time.

Krisen Moodley, BSc, MSc

Krisen Moodley is Director for Postgraduate Taught courses in the School of Civil Engineering, University of Leeds. After graduating from the University of Natal, his initial employment was as a quantity surveyor with Davis Langdon Farrow Laing in Southern Africa, before his first academic appointment at Heriot-Watt University. He spent four years at Heriot-Watt before joining Leeds in 1994. His research interests are concerned with the strategic business relationships between organizations and their projects. Other specialist research interests include procurement, project management and corporate responsibility. Krisen has published and presented papers on these subjects both nationally and internationally. His recent books include *Corporate Communications in Construction; Construction Business Development; Meeting New Challenges, Seeking Opportunity*. He also has contributed chapters to *Construction Reports 1944–1998, Engineering Project Management* and *Commercial Management: Defining the Discipline.*

Dr Christopher N. Preece BSc (Hons), PhD, FCIOB, MCIM, ILTM

Dr Christopher Preece is a Fellow of the Chartered Institute of Building and a Member of the Chartered Institute of Marketing. He is a Member of the Engineering Professor's Council and Royal Academy of Engineering Group on the teaching of engineering ethics. He is a Lecturer in the Construction Management Group of the School of Civil Engineering, University of Leeds, where his lecturing and research work centres around the application of business ethics and corporate social responsibility in the built environment and engineering sectors.

Acknowledgements

We would like to thank the Institution of Civil Engineers, the American Society of Civil Engineers, the Engineering Council, the Royal Academy of Engineers, the Institute of Electrical and Electronic Engineers, the Institute of Engineers Australia and the California Contractors and the many other institutions mentioned in this book for access to their ethical codes and codes of practice.

1 *Why be ethical, or whose responsibility is it anyway?*

> Once the rockets go up, who cares where they come down?
> That's not my department says Werner von Braun
>
> Tom Lehrer.

Introduction

At the outset we offer you working definitions of some key terms. These definitions will evolve throughout the book:

Ethics: The philosophical study of what is right or wrong in human conduct and what rules or principles should govern it. Hence the term is singular. This is often subdivided into meta-ethics, applied ethics and professional ethics.

Meta-ethics: The systematic study of the nature of ethics. This looks into issues such as how an ethical judgement can be justified and the possible theoretical underpinning of ethical reflection and practice.

Applied Ethics: The application of ethics in a particular area of practice, e.g. business or bio-ethics.

Professional Ethics: The ethical identity, codes and practices of particular professions, such as the professions followed by nurses, doctors, lawyers or engineers.

Morality: Morality often refers to standards of moral conduct – right behaviour. In the history of philosophy there have been many attempts to differentiate the concept from ethics. However, it is most often used interchangeably with the term 'ethics', which is how we will use it.

Engineering Ethics: Engineering ethics is defined in the two ways:

1. The study of moral issues and decisions confronting individuals and organizations involved in engineering.
2. The study of related questions about moral conduct, character and relationship involved in technical development (Martin and Schinzinger 1989).

Our study will focus on the professional and business ethics of the engineer, but will also make reference to other areas, not least through broader debates. As we explore ethics it will become apparent that whilst it is possible to see the study of ethics and its theories as a separate discipline, ethics is at the heart of any significant decision-making, and is therefore central to reflective professional practice. Often we make a statement without realizing that it involves particular values or principles. Often only in moments of challenge do we begin to state what those values are.

Perceptions about ethics can be affected by negative stereotypes. It is frequently seen as essentially about philosophical theory, and therefore seen by many as not relevant to practice. Hence, as Vardy (1989, 194) notes, practitioners are often suspicious of the philosopher, whom they see as living in 'a secure and problem-free environment removed from business realities'. Ethics can also be seen as prescriptive and judgemental. Seemingly embodied in rules and codes, it is viewed by some as imposed from above and therefore against the idea of the freedom of the individual to make his or her own ethical decision, i.e. against ethical autonomy.

We will argue for ethics as essentially interdisciplinary. In this, philosophy and the social sciences have important parts to play in:

- locating and connecting underlying belief systems to individual and corporate ethical practice,
- developing shared discourse in value and belief systems,
- developing the tools of critical thinking, and
- enabling critical reflection on and dialogue about underlying ethical theory and meaning.

Importantly, no single discipline can 'own' ethics. It is focused in reflective practice, and as such, any discipline is there to inform and support that practice. We will also argue that ethics underpins many different concepts that are often seen as quite distinct, from sustainability, to corporate social responsibility (CSR), to corporate governance, to professional virtues and moral awareness.

In this chapter we aim to do two things:

1. To explore the reasons for being ethical.
2. To examine the nature and purpose of the engineer, and how a concern for ethics emerges from these.

Case 1.1

A memo from engineer Willy Just: A shorter, fully loaded truck can operate much more quickly. A shortening of the rear compartment will not disadvantageously affect the weight balance, by overloading the front axle, because a correction in the weight distribution takes place automatically through the fact that the cargo in the struggle towards the back door during the operation is preponderantly located there. Because the connecting pipe is quickly rusted through the fluids, the gas should be introduced from above, not below. To facilitate cleaning, an eight- to twelve-inch hole should be made in the floor and provided with a cover that can be opened from the outside. The floor should be slightly inclined, and the cover equipped with a small sieve. Thus all fluids will flow to the middle, the thin fluids will exit during the operation, and the thicker fluids can be hosed out afterwards. (Bauman 1989, 197)

Just had been commissioned to improve the efficiency of the trucks developed by the Nazi regime in the early 1940s to transport and gas Jews and others. It may seem an extreme example of engineering, and yet there were many professionals in Germany at the time faced with the dilemma of having to respond to such commissions. On the face of it Just had very good reasons for not being 'ethical'. If he had opposed his clients then his and his family's lives would have been in danger.

Another school of thought, though, might suggest that 'being ethical' did not come into it. Just was an engineer and the task of engineers is to do what the client or manager wants. The issue of ethics, of how any commission is put to use, of where the new project is built, who it may affect and so on can seem like concerns solely for the client, the local planners or the government to consider. These are the parties who commission the project and who take responsibility for the ethical issues. It is their duty to be aware of potential problems, value conflicts and so on.

This view could be further reinforced with an argument that the professional engineer should be entirely impartial about 'values'. His or her task is simply to complete the job to the highest technical standards and not be interested in, or influenced by, the values of the client or any others who have a stake in the problem – many of whom are probably not equipped to understand the 'engineering' in any case! But of course this does not happen. Something stops the engineer from 'building anything for anybody' without considering consequences. The client is similarly constrained. Both fall back on the 'ethics of the situation' – often without being able to explain what this means, but either or both could perhaps, in a given situation, justify making small but significant changes to their 'ethical' position.

So we cannot get far in these considerations without asking what it means to be 'ethical' as a professional engineer or manager. It could be argued that Just actually was 'ethical'. Whilst intuitively this seems wrong, there is no doubt that the Nazi regime was built on a self-generated value system, underpinned by a complex belief system. Burleigh (2000) argues that Hitler aimed to provide a replacement religion to inspire his followers. We may not like the Third Reich ethical system; indeed, we may argue that it was profoundly evil, but it had a contemporary view of what was right and wrong in that country. Just may have accepted it or believed in it in the same way that many SS guards interviewed several decades after the war still did (Rees 2005, 132 ff.). This argument can be refined to suggest that there is no one view of ethics and that all ethics are relative to their social and cultural context.

From another perspective, we do not even have to accept relativity to argue that Just was in fact behaving ethically. He was in an impossible situation, unable to refuse the commission. In that light the most ethical thing to do was perhaps to ensure that his family and his workforce were not harmed. It was the best of several bad options.

A nagging doubt, however, persists. Didn't Just, as an engineer, collude with countless others who chose to turn away when Jews were killed in the streets, not to 'see' the crematoria close to their villages, or not to challenge the bullying 'brown shirts' who whipped up hatred? Whatever the stance taken, however, the memo makes chilling reading, not least through the way in which technical language is used such that it denies the humanity, and with that the suffering, of those at its centre.

The argument can ask a further uncomfortable question. How did the personal ethics of Just relate to his professional ethics? Shouldn't there be some consideration about a view of what is 'right' that transcends any limited view of the situation, or of the profession, or any contract, consideration of obligations to humanity as a whole?

A way of looking at this argument is to suggest that such a view is actually part of what it means to be a professional. Engineers and managers operate in society and beyond, and cannot distance themselves from that broader relationship, or from the issues that surround them. 'Knowledge is obligation' and the knowledge and power of the professional engineer and manager must be matched by their sense of responsibility.

None of this is to condemn Willy Just. The reflection on his memo is simply to tease out an answer to the question, 'why be ethical?'

Exercise

1. Put yourself in the position of Willy Just. How would you have responded to the commission? How would you have justified your response?
2. Work with two other colleagues and imagine that you are the board of Just's firm. How would you respond together? What would you do as a group in that situation?

On thing is for sure. In a very short time, our reflection on 'why be ethical?' has moved to the question 'what is ethics?', and our next step is the question 'is ethics about right and wrong in a relatively narrow context or does it reach out to involve the professional relationship with wider society?'

A second case involving software engineers might help us to dig a little deeper.

Case 1.2

Following the success of a computer game based upon a horror scenario set in the frozen north the computer software development company was asked by the client to develop a second game. This time the client wanted increased shock value, and the inclusion of the explicit death of young children. An added incentive would be that agreement to the commission would lead to the rapid release of monies still owed to software firm for the first game.

The manager of the software firm and his engineering staff were uneasy about this request – though initially a little unsure why they felt this unease. As a result of discussions with his staff the manager decided that it was important to clarify the situation. He wrote to his client's legal department and asked if they would confirm in writing that the company wished him to develop a second game and that it was their intention that this should involve increased horror and the death of children. No such confirmation was received and the money owed to the software development company was rapidly released.

This case is interesting at several levels, not least because effective resolution seems to have been made by the client without anybody reflecting on and articulating any ethical meaning about what was being proposed. The computer engineers shared a strong feeling of unease, but at no point was the meaning behind the unease spelled out. The response of the legal department could be surmised as self-interest, but was not spelled out.

This case has been used at different times with students and professionals as a method of exploring ethical standpoints. Without the participants having the benefit of knowing how the software manager had replied, there were a wide variety of responses to the situation, including the following:

- One group suggested the most important thing was to give the client what they wanted. To this end they determined to make the children Eskimos, i.e. natives of the land in which the game was set. No one in the West, they argued, would identify with Eskimos, and so there would be no offence caused.
- A second group suggested that games have their own context, and this is not real. Nothing of what happens in a game can affect reality.

- A third group argued squarely that having a game involving the death of children was not ethically wrong, provided there was an ethical frame to the game, such as the killer being brought to book.
- A fourth group suggested that the issue of what was happening should not be the responsibility of the software engineer. The company had commissioned the game and it followed that it was their responsibility to deal with any ethical issues that it generated, not least the issue of how computer games affect the character of the players (Reber 2001).
- A fifth group suggested that to go with the commission was too risky. It only needed one press release that inferred that the client company was trading in on the death of children, and whatever the ethical rights and wrongs of this, it could adversely affect the company's reputation and thus sales. This could in turn affect the reputation of the software engineer.

Exercise

1. How do you see Case 1.2?
2. How would you work through the feelings of unease felt by the software company?
3. How would you respond to the client, as an individual or as the manager?
4. Imagine that you are the manager of a small family-owned, medium-sized company. The family board members have called you in. They are relieved that you have dealt successfully with an issue similar to that described above. They are keen to ensure that the firm learns from the experience. They want you and the other senior software engineers to identify the key issues and to develop a policy for how future similar problems might be approached.
5. Working with two colleagues or students, write a response to the board.
6. Show this to two others, aiming to justify your background thinking and approach to them.

Case 1.2 and the various responses seem to indicate two answers to the question, 'why be ethical?' First, it is in the engineer's interest to be ethical. Second, it is a conditional answer. The engineer should be ethical if he or she has a clear responsibility.

The ethics of self-interest

Ethics can involve self-interest:

- A sense of ethical awareness gives an engineering organization an image that encourages the public and clients to trust them, reinforcing the importance of an ethical reputation. Though no reason was given by the client in Case 1.2 for not pressing the original brief it is safe to assume that a company that had a strong interest in the family market would not want such a brief to be in writing. One of the most interesting examples of self-interested ethics in business in general is the case of Nestlé and the sale of a baby milk substitute (Robinson 2002). This case has raged for over thirty years. In its earliest stages the whistle was blown by *The New Internationalist* on the alleged practice of providing free baby milk substitute to poor third world mothers in hospitals, leading to dependency upon the substitute and eventually to the deaths of millions of babies due to poor mothers mixing the powdered milk in infected water. Whatever the rights or wrongs of this complex case, there is little doubt that Nestlé's initial handling of it was ill judged. Rather than taking these claims seriously, and trying to address the issues and the underlying value conflicts, Nestlé tried to paint them as anti-capitalist propaganda, taking on the role of champions of the free market. First, this led to a polarization of the issues. Then it led directly to an economic boycott of all Nestlé goods. Only when Nestlé entered into serious dialogue did the issue become manageable. Since that time Nestlé have been working hard to salvage their reputation, resulting in significant changes, including the development of a World Health Organization Code on marketing.
- It is in the interest of the engineer and all who are affected by particular projects to be ethically aware. The ultimate consequence of not dealing with core ethical issues, such as health and safety or environmental concerns, can be major disasters, ranging from Bhopal to Five Mile Island. Such disasters reinforce the central point that engineering work can radically affect different stakeholders (defined as those who are affected in some way by the industry) in ways that have far more serious consequences than the average business failure. Without attention to the interests of diverse groups the engineering firm is unlikely to make progress with the clients who are important to its work.
- Awareness of ethical issues and the capacity to handle them before they become a problem will enable the industry/profession to avoid government regulation and legal sanctions.

For the most part it is fairly obvious when legal considerations demand ethical behaviour. In the late 1980s an engineer employed by a steel firm in Sheffield was invited to meet a foreign client in Manchester airport VIP lounge.

The 'client' turned out to be four generals from Iraq and the commission was for a series of 'long pipes', very precisely specified and with a very large diameter. When the engineer questioned the purpose of the pipes he was told that they were for an oil pipeline. When he asked about the involvement of generals in the project he was told that all engineers in Iraq were in the army. In this case the decision of the engineer was fairly clear. Although the meeting took place before the Iraq supergun scandal came to light, he realized what the purpose of the 'pipes' might really be and knew that the UK law prohibited involvement. He was in any case not happy about being involved in the production of weapons of war.

It is interesting, however, to compare this engineer's reactions to the project with the way other sections of UK industry reacted to similar requests from Iraq at later dates. In April 1990 Customs officials at Teesport, UK, seized eight steel tubes that were made in Sheffield and destined for Iraq. A Sheffield company had received an order for two sets of twenty-six tubes of 1000-mm-diameter constant-bore. Joined together these would form two 156-m-long smooth bore pipes. (By way of comparison the World War 1 howitzer 'Big Bertha' was 420 mm in diameter while Krupp's World War II K12 gun was 210 mm by 36 m long.) The eight tubes seized at Teesport were the final eight; forty-four had already been shipped. Some of these were stopped en route to Iraq in Greece and Turkey. As well as seizing these pipes, Customs officials also visited a second British firm involved in the project who had supplied Iraq with a series of 350-mm-diameter tubes together with a hydraulic mechanism. The drawings for these were supplied by a company based in Athens and controlled by a space research company based in Belgium.

A later report claimed that at least six Britons had worked on the project. Some of them had answered an advertisement placed in a Bristol newspaper for 'Design Draughtsmen' at enhanced rates of pay. It was also revealed that the UK Ministry of Defence experts had been aware of the order and had looked at the tubes and documents of the firms involved. These experts considered that the 'indications were that the tubes were components of a large-calibre armament, albeit of a scale outside anything previously experienced.' The UK secretary of state confirmed that his department had been approached by one of the firms to ask if a licence was necessary for the export of metal tubes to the space research company in Belgium.

The Sheffield company had made a similar enquiry in relation to the export of tubes to Iraq for use in the polymerization of polyethylene. The secretary of state reported: 'On the information available at that time it was decided that export licences were not needed for those goods. Until a few days ago, my Department had no knowledge that the goods were designed to form part of a gun. If my Department had known that purpose, then it would of course have advised that licences were necessary, and they would not have been granted.'

Information tends to suggest that the UK government knew that Iraqi military personnel were involved in the order and still the advice given to the two firms was that they did not need export licences. The then Prime Minister Margaret Thatcher was adamant that the blame lay with the firms and that it was up to them to seek export licences. As time passed, revelations about the 'supergun' multiplied. Additional 'parts' of the gun were reported from Italy, Switzerland, West Germany, and Spain. Belgian ammunition was being developed as the propellant for the gun and at least one cargo was said to have been shipped via Jordan.

The passage of time has still not provided a final version of the 'truth' of this case. It is still not really known whether there actually was a 'supergun'. Questions regarding why Iraq would want such a gun when the missiles it already possessed could do the same job of delivering warheads over long distances much more effectively are still asked. The gun was to have been so large, to have such a fixed position and be so difficult to conceal so as to make it an immediate target for any enemy. All along Iraq has claimed that the 'supergun' was piping for its petroleum industry. The then Foreign Minister Tariq Aziz was quoted as saying, 'Even if we buy a box of chocolates from Britain, they'll say Iraq will use it to produce an atomic bomb.'

Customs and Excise eventually dropped all the charges brought earlier against the two firms in connection with the 'supergun' affair. The prosecution decided there was not a realistic chance of proving to a jury that the directors knew how the exported pipes would be used and Customs decided it would be 'inappropriate' to continue to press charges (www.caat.org.uk 2002).

At its most basic it could be argued that in the first case the Sheffield engineer was obeying the law, this providing a built-in reason for being ethical. In the second case the ethics surrounding the situation of the two firms is more complex. Advice and guidance sought from government sources initially suggested that they had done nothing against the law but this decision was later reversed and attempts were made to prosecute them. It is hard to imagine, however, that all involved did not suspect that something unethical was being done despite the initial reassurance that their actions were not illegal.

Some may argue that ethics is the same as the law, and there is no doubt that it makes sense to obey the law. However, whilst good laws are built on moral values, the law is not necessarily the same as ethics. Ethical judgements can, and often do, challenge the law. Changes in the ethical thinking in the 1960s, for instance, resulted in suicide and homosexuality no longer being regarded as criminal offences. In engineering the law often reinforces basic ethical imperatives, not least in the fields of health and safety, but there are often situations where the law does not apply neatly, especially when it comes to questions of responsibility, as we shall see in the case of the Challenger.

So whilst the law is a good reason for being ethical, it is not the whole reason. In a global perspective there are many different legislative approaches, and the relationship of law to ethics might be questionable. The law with respect to the environment or employment in some parts of the world, for instance, might not take account of environmental, safety or human rights issues.

Arguments for an ethics based on self- and mutual interest, then, are reasonable, and indeed necessary in the modern marketplace. It would be difficult to argue the case for an ethical stance that would result in the demise of an engineering business. Underlying much of these arguments is the importance of trust in the profession. The effective practice of the profession depends upon trust. Hence, the individual professional holds a responsibility not simply to the client or to the wider society but to the profession, such that trust in the profession as a whole can be maintained. This is partly trust based upon the knowledge that the professional will maintain the highest possible technical standards. The professional has to be competent. However, the competence has to extend beyond the technical to the relational, and to the belief that the professional will not take advantage of, and also seek the best for, clients who are, in many cases, vulnerable.

Who is responsible?

The clear answer from some of the student responses to Case 1.2 was that for the most part the client was responsible. The engineer is then 'let off the hook' and does not need to worry about the ethical issues.

Samuel Florman (1976) would argue that this can provide the basis for a limited view of why an engineer should be ethical, and what his or her responsibilities are. He argues that the core responsibility of the engineer is to the employer or client. The professional engineer has the task of serving them and meeting their expectations, rather than filtering everything through a broad conscience. If that relationship is properly fulfilled then the only constraints on the engineer are compliance with government regulation and laws. Remember this remains an ethical position and is summed up in Florman's (1976, 32) words as: 'Engineers are obliged to bring integrity and competence to whatever work they do. But they should not be counted upon to consider paramount the welfare of the human race.'

Florman's position is interesting but troubling. Firstly, it is not clear what he means by integrity. This will be examined more fully in Chapter 2. At this point it is sufficient to note that integrity might mean consistency between personal and professional values. If that is one aspect of this concept then it is difficult to see how one can restrict the ethical domain in the way that Florman tries. The two supergun cases show how the engineer is faced by personal and public values that are greater and more pressing than the immediate relationship to the client. Moreover, the second of these cases shows that the law is not always clear. Secondly, Florman posits that there is one major relationship upon which the engineer should focus and this is the basis of the responsibility of the engineer. He provides no reason why one should narrow the ethical responsibility of the engineer in this way only and no reason why the professional engineer should not have several responsibilities. Even an ethics based on self- and mutual interest begins to point to multiple responsibilities, which in turn provide a more coherent basis for why the engineer should be ethical.

This focus on responsibilities begins to move the debate beyond a simple self-interest argument for being ethical. To explore this argument we will examine the case history of the Challenger. We make no apologies for using this case history, already well explored, not least because it highlights issues about the identity of the engineer.

The Challenger 51-L case history

On 28 January 1986, only 73 seconds after launch the Challenger Space Shuttle exploded, creating one of the most public and high-profile disasters of the twentieth century.

Was the failure of the Challenger an ethical or technical failure – or both? At the heart of the case were the central engineering design problems focusing on the field joints of the external fuel tanks. These were sealed by O-rings. Before the launch of the Challenger 51-L there had been consistent problems with the O-rings, with clear evidence of gas blowing through the seal. The engineers involved continued to work on developing the design of the joint, but by the time of the flight in January 1986 the redesigned joint was not available.

The social, political and economic context of the engineering work was complex.

- NASA was under intense pressure, both from unexpected external competition and from the US government. It needed to make the strongest case for the Space Shuttle programme and budget and had scheduled a record number of missions for 1986. The future of the NASA programme depended upon accommodating all these.
- There was also additional pressure around the US intention of beating the Russians to probe Halley's comet. This would require the 51-L to meet rigorous programme dates and to be refurbished rapidly.
- Political pressure was added by the White House, which wished to ensure that the 51-L was in orbit during the State of the Union speech. President Reagan was going to make specific reference in this speech to Christa McAuliffe, the first civilian and the first 'ordinary hero in space'.

The pressures on the engineers were intense, not just around the launch of 51-L but throughout the whole programme. From the beginning of the programme in 1981 the issue of safety was not addressed with any consistency. Documentation of safety issues was not well maintained, and there was no independent scrutiny of the safety systems. In fact, safety was viewed as a matter of increasingly low priority. This was exemplified by the idea of an escape system for the crew, should the mission be aborted soon after take-off. This initial proposal was omitted after consideration, in order to maximize the fuel payload.

As plans for the launch were made, ongoing problems with O-ring seal erosion were raised by the engineers of Morton-Thiokol Industries (MTI). Morton-Thiokol Industries had been awarded the contract to build the solid fuel rocket boosters for NASA as far back as 1972.

A task team was set up to address the problems, but did not work well. Far from providing a speedy resolution, there were further delays. Ebeling and Boisjoly, the two main MTI engineers on the team, wrote that management as a whole did not understand the importance of the task and that NASA bureaucracy led to constant delays. At one point requests for spare parts had to go through eight different offices. The last report sent by Boisjoly, highlighting the problems leading to delays, received no reply from his own or NASA management.

Boisjoly kept good records of his interaction with management, and was clear about the possible consequences of ignoring the O-ring problem. He ensured that all memos were circulated to his direct superiors. However, he and his colleagues were not prepared for the less than positive response they received from management. Towards the end of the team's work, when they had not received feedback, they did not press their concerns or seek to clarify whether their managers or NASA had understood the potential consequences. Underlying the work of the task team was the fundamental issue that they were dealing with a design fault and that the fault was so potentially serious that it should have led to the recommendation that the shuttle programme be suspended until it was corrected. NASA engineers were aware that there was a problem and for a time worked with the task team. However, despite the technical concerns the problem was still viewed and defined by management in terms of trying to speed up the delivery of the new design rather than as an essential matter of safety. One of the difficulties was an assumption that the response of the NASA engineers seemed to indicate to MTI that NASA had heard and understood Boisjoly's complaints. In fact, the presence of the NASA engineers on the task team did not mean that the issues raised were being communicated to the NASA managers. Indeed, the essential matter of safety remained unacknowledged by NASA management, who even claimed to the Presidential Commission, set up after the disaster, that they were not aware of the problems.

Throughout the build-up to the launch the MTI engineers operated strictly within the established communication system, reporting to their managers and relying on their managers to communicate the problems to NASA.

The problems in the established system of dissemination of information meant that the MTI engineers' concern was never full communicated, never fully appreciated by their own managers and never adequately passed on to their client NASA.

Questions

What more could the MTI engineers have done at this point?
How might the MTI management have best handled the material coming from their engineers?

How could NASA have kept itself better informed about this aspect of the project safety?

All those involved in the shuttle programme were aware of its high profile and the consequent pressure to achieve targets. The result was a strong incentive to think and plan unrealistically. As far back as 1982 NASA began a planned acceleration of the launch schedule. An early plan anticipated an eventual launch rate of one per week. By 1985 this had been scaled down to a projection of an annual launch rate of 24 by 1990. Realistically, even this modified goal was overambitious leading to strain, including:

- A critical shortage of spare parts.
- Strain on the IT production system, which meant that it would not have been able to deliver to the crew training software for scheduled flights by due date. This in turn would have meant inadequate time for crew training.
- No enforcement of cargo manifest policies, leading to numerous payload changes at the last minute.

The Presidential Commission underlined how this affected attention to safety:

> When flights come in rapid succession, current requirements do not ensure that critical anomalies occurring in one flight are identified and addressed appropriately before the next flight.
>
> (Report of the Presidential Commission (RPC) 1986, 14)

The situation came to a head in a teleconference, organized as late as the night before the launch. The aim of the conference was to clarify the conditions for launch safety. Before the conference, MTI's position was that 'the condition is not desirable but is acceptable' (RPC 1986, 12).

The result of the teleconference was, in the words of Commission member Feynman, a 'kind of Russian roulette' in which standards of safety were gradually lowered. It was argued that if the shuttle had flown with O-ring erosion and nothing happened, 'then it is suggested that the risk is no longer so high for the next flights' (RPC appendix E3).

None of the discussions were based on thoroughly researched evidence. The true problem was not really perceived by MTI managers or by NASA, and no reliable background information on which to make informed decisions was available. As the subsequent Presidential Commission found, before the accident neither NASA nor MTI managers or engineers fully understood the mechanism by which the field joint operated and the joint was sealed.

Whilst safety was the issue for which the teleconference was convened the Commission discovered a massive discrepancy between the engineer's and the management's view of safety margins. Engineers estimated a shuttle failure rate of 1 in 100 launches. NASA management had figures of 1 in 100 000. When the NASA figures were questioned by the Commission the response of their chief engineer was, 'We did not use them as a management tool. We knew that the possibility of failure was always sitting there . . . ' (Martin and Schinzinger 1989, 83). This leaves the unanswered question as to how each figure was determined. As to the use of such figures, it seems that they were submitted in response to a risk analysis for the Department of Energy. The true aim of the risk analysis was to calculate safety in the use of small atomic reactors as power sources for deep space probes the shuttle could carry into space. The next mission of the Challenger was scheduled to carry the Galileo probe, with 47.6 pounds of plutonium-238.

The NASA figures were in Feynman's words exaggerated 'to the point of fantasy', and were part of a dynamic that led to a reversal of the usual view of safety. NASA management even argued that the fact that a third of the O-ring had been eroded and the shuttle still continued to fly demonstrated a 'safety factor of three'. No one involved in the Commission could understand what this phrase meant, given that erosion of a seal indicates a diminution of safety by a third, not a safety margin of three.

All this confusion was brought as a background into the teleconference, which focused on the question of the performance of the O-ring at the ambient temperature at which the launch would take place. Although there was clear correlation between O-ring damage and low ambient temperature, neither NASA nor MTI had carried extensive tests at temperatures below 53 F. The engineers had not been allowed the time or resources to research these effects and to prove conclusively that it would be safe or unsafe to fly at these temperatures, precisely because this additional programme time would have resulted in the shuttle programme being suspended. Hence, NASA and MTI did not have the requisite information and were not adequately prepared to evaluate the risks of the launch in conditions more severe than had been experienced before. The engineers all strongly argued that on the basis of the initial – if incomplete – temperature correlation the flight should be delayed, but they did not have information from tests carried out at all relevant temperatures.

The response of the NASA and the MTI managers to their argument was threefold. Firstly, there was hostility, doubtless because of the prospects of a further series of delays and because of the pressure they were experiencing. Secondly, they questioned the engineer's figures. NASA engineers questioned the temperature figures, believing that the original design specification they had provided was for temperatures as low as 31 F. The extent of confusion was to emerge later when MTI argued that the temperature of 31 F was the limit for storage and not for operation. In any case the launch temperature was below this figure. The NASA managers focused on the lack of absolute evidence regarding safety and on the fact that previous O-rings had only been eroded by a third, without compromising safety. Finally, it was later argued that the MTI engineers were not unanimous in their judgements.

The MTI engineers found themselves suddenly having to prove that the Challenger was unsafe to fly at low temperatures – without the real evidence to prove their case. Hence, the basis of the professional judgement had moved from true risk assessment to the need to prove that it could not be launched in these conditions with unsubstantiated evidence. In the words of Lund, another of the MTI engineers, in his testimony to the Commission:

> We [had] . . . always been in the position of defending our position to make sure that we were ready to fly, and I guess I did not realize until after that meeting and after several days, that we had absolutely changed our position. . . . And so we got ourselves into the thought process, we were trying to find some way to prove to them it wouldn't work, and we were unable to do that.
>
> (Harris et al. 1995, 285–6)

There was a break in the teleconference during which all groups concerned reviewed their positions. The situation now highlighted major questions of responsibility, opening up differences of perspective between engineers and management. Lund was both an engineer and a manager and he was, at that moment, invited to wear his 'management hat'. This implied very different responsibilities. Harris et al. (274) suggest that engineers are more traditionally concerned with risk–benefit analysis and management with cost–benefit analysis. Lund was in effect invited to put aside the questions of risk raised by the MTI engineers, to think primarily of cost–benefit as a manager. The Presidential Commission was more direct, concluding that MTI had altered its original decision, 'to accommodate a major customer' (RPC 11).

Whatever the final judgement, it is clear that managers and engineers had come to a critical point in decision-making, without any prior agreement not only on safety procedures, but also on the criteria of standards for measuring safety in the first place. By being drawn into the management issues, engineers were deflected from their natural defence of the key principle of safety. After further discussions, on returning to the teleconference the result was a decision to launch.

Questions

Once this responsibility divide was crossed the decision was effectively taken away from the engineers. Whose responsibility then was it to maintain safety? What could the MTI engineers have done differently in this situation? Was their disagreement fully communicated to NASA? Is it morally acceptable for an engineer or other employee to make a case about safety and then pass the responsibility for the outcome to the management involved? How would you have responded in that situation and how would you have justified your response?

The impression received in this case is that once the teleconference had ended the die was cast. However, managers and engineers could have intervened up to the point of the final minutes of countdown, simply through direct communication with the NASA head of launch, Arnold Aldrich. He expressly encouraged any key personnel to contact him at any time before the launch.

The problem was that the management system worked against such open communication. The decision-making process was very complex and fragmented, with different sites and different levels of decision-making. Lawrence Mulloy, for instance, represented Marshall Space Flight Centre at the Kennedy Space Centre. He therefore acted as the conduit for information to NASA launch staff. However, he did not tell Arnold Aldrich the details of the teleconference because he was at level III and Aldrich at level II of decision-making. This encouraged those at level III to solve problems and make decisions, without sharing the full background to these decisions with level II and I personnel. Hence, the Presidential Commission was troubled 'by what appears to be a propensity of management . . . to contain potentially serious problems and to attempt to resolve them internally rather communicate them forward' (RPC 11). In effect, the management system encouraged narrow thinking, concentrating purely on a limited area of responsibility without seeing the broader connections, and also secretive thinking, not sharing difficulties that might cause problems or require major work for the next level of management. This in turn led to a lack of transparency and with it a lack of concern about critical safety procedures with which higher management were not kept abreast. Hence, for instance, there was no system that ensured that launch constraints and waivers of launch constraints should be considered at all levels of management. This led to the six launch constraints (including adverse weather) being waived before the launch of 51-L, with no record of the constraints or of the waivers.

The fragmentation of decision-making led the different parts of the organization to focus purely on their particular areas of responsibility, and not to review or feel responsible for safety as a whole. The Presidential Commission and subsequent reports stressed that all levels of management should feel responsible for safety, and not simply leave it to one designated group.

Questions

What difference does the management perspective make to the engineer?

What recommendation about future practice would you have made to the board of MTI?

The move from the engineer's to the manager's perspective begins to crystallize the identity of the engineer in this case study. Up to that point the perspective of the MTI engineers was radically different from that of the managers, and from that of members of other firms. This case history can help us consider what is distinctive about the professional engineer and how the very nature of the engineer can furnish us with greater clarity about the responsibility of the engineer and reasons for being ethical that take us beyond self-interest. We will begin with the engineer as a professional.

The engineer as a professional

The concept of a professional tends to be an elevated one. The following features can be associated with professionals:

- Specialized knowledge and skills.
- Power – the power of specialized knowledge and the capacity to significantly affect others, be they persons, groups or the environment.
- A monopoly or near monopoly of a particular skill.
- Managers undergo an extensive period of training that includes the development of both skills and the intellect.
- Membership of a professional body that is responsible for maintaining standards, protecting rights and ensuring proper training.
- Autonomy of practice.

The engineer clearly falls under this definition of profession, and analysis of some of these points reveals core values.

Professional autonomy

Professional autonomy here is a concern for the professional to not be compromised or partial. This is about a negative freedom that ensures that the professional is not influenced to move into a 'technical judgement' made on the basis of interests of other parties. Once the engineers in the Challenger case moved away from their professional autonomy into a more managerial perspective, their decision was compromised. Also, by bracketing professional judgement, the underlying technological and scientific truth was corrupted and managerial rather than scientific criteria were used to justify the launch. This is a double cautionary tale for the professional: lose your professional autonomy, and professional and scientific integrity can soon follow, with potentially disastrous results. At the heart of this is a moral imperative about providing the proper technical data, such that the implication of the technical proposal is clear.

Of course, there are limits to professional autonomy in that the profession has to deal with different constraints. This is felt even in medicine. The doctor is not simply practising in relation to the patient, he or she also has to satisfy other 'clients', not least the government, by improving a service within financial constraints. Nonetheless, if the doctor or engineer were to be influenced solely by such perspectives this could affect the competent delivery of his or her services. Put bluntly, the responsibility of the professional engineer in all situations is to tell the truth, based upon the facts of those situations. In effect, this involves the freedom to tell the truth to the client. The client or employer has to make key decisions and this requires that he or she be fully informed.

The client/employer relationship

Florman's view of the engineer as owing responsibility primarily to the client is simplistic. It first assumes that the client knows exactly what he or she wants. However, this may be far from the case. The client/employer may have an overall idea, in which case the engineer has to help set that idea into technical reality, by including any constraints that impinge upon the idea. This will in effect lead to an ongoing conversation about aims and objectives and what the technical response might involve. In some situations this might be a very short discussion. In others the complexity of the physical and social environment may make it protracted. Hence, the task is not simply giving the client what he or she wants.

This affects the autonomy (self-governance) of the client. The Florman view assumes that the client/employer not only has made his or her decision but that it is autonomous and that this autonomy should be respected. This is a limited view of client autonomy and implies that their advisers are expected to effectively help them to achieve anything that they wish. A more realistic view of client/employer autonomy is that they are only fully autonomous when they are made fully aware of the range of technical issues and how they impinge on the social and physical environment. Behind this is a moral imperative of the engineers to not simply communicate the full technical truth but also ensure that the decision of the client is fully informed. This runs parallel to the medical principle of informed consent. In the Challenger case it is clear that the clients were not fully informed. They had not worked through the implications of what they were aiming to do.

On the basis of this thinking, Koehn (1994) and May (1985) suggest refining the definition of a profession in terms of the professional empowering the client. From this perspective, client autonomy and responsibility are seen as developing as part of the professional relationship, and a great deal of the professional's role is about teaching, clarifying and enabling. All of this implies a responsibility of the engineer not simply to communicate the truth but also to ensure that the client has 'registered', understood and accepted the truth, i.e. the responsibility of the engineer is not just to communicate the data but to argue the case in a manner the client can understand. In Case 1.2 the letter to the legal department partly had this effect by asking the client to clarify what they wanted. The very process of clarification, we assume, had the effect of making clear the potential ethical implications. The practice of one engineer is to put in writing in the report all the implications to society and environment and have the client sign that he or she has fully understood.

The Challenger case shows how many things can get in the way of successful communication. The different levels of management led to discussions that were sealed off from higher managers. The pressures from different groups were such that the technical truth was distorted and overruled by the truth perceived by the management.

There could be cases where what is being suggested by the potential client is simply wrong, and the Challenger case could be said to fall into this category. If the client, with whom the engineer may have a good and lucrative relationship, is suggesting something that is simply wrong how is the engineer to respond? In the Challenger case one could argue that the demand to launch was not against the law, but was straightforwardly technically and ethically wrong. If clarification and dialogue do not affect the client's decision then the engineer has to make a judgement. It could be argued that not to challenge the client's decision would be to collude in the wrong act. The key to successful challenge is repetition and transparency. Repetition shows that the engineer's judgement is not going away. Transparency means that others may see what the client is responsible for. The extreme of such an action is whistleblowing.

None of this is to say that the engineer has sole responsibility for ensuring that his judgement is properly taken account of. That responsibility is shared with the client or the employer. It is in his or her interest to value the engineer's professional autonomy and the enabling role, precisely because it will ensure that the client is made clearly aware of all the facts and the consequences of different options. For any firm this also means that its system of management should be transparent, enabling clarity and good communication.

Shared responsibility for the project

All of this suggests that the relationship of the client/employer to the engineer can be complex. In the Challenger case there were in fact many different clients/employers, from MTI, the different companies and contractors involved, to NASA, to the government. Each had responsibilities to the other and lines of communication were not simple.

At one level this meant that the MTI engineers had to deal not only with recalcitrant responses from different management groups, but also with different expectations and demands from different clients. More particularly, this suggests that it is important not to see the engineer's responsibility to the client in a restricted way, with the engineer simply doing his or her job and then passing information to the client, who then becomes responsible for the final decision. On the contrary, there is a sense in which both the client and the engineer *share* ethical responsibility, as they work through dialogue about the decision or project. This is a conclusion that the Presidential Commission seems to endorse. They suggest that the problem with fragmented management structures is that they also fragment any sense of responsibility for the whole outcome. This kind of thinking, of course, is reinforced by narrow views of responsibility, which associate it with liability. Hence, for many people there is the desire to avoid such a view of responsibility. For example, in another context there were several crew members on the Herald of Free Enterprise ferry who saw that the bow doors were open and did nothing about it (Robinson 1992). The ferry sank with great loss of life. The crew members explained their action by saying that it was not their job or their responsibility.

The Presidential Commission argued that everyone, including the engineers, should feel responsible for the outcome of the project. Once again, this requires, but is much more than, good lines of communication and dialogue. Of course, few engineers will have to face the complexity and tension of the Challenger situation, but the principle of shared responsibility for the outcome of a project remains at whatever level engineering works.

Shared responsibilities to others

The engineer does not operate in a vacuum. He or she is involved with different kinds of power, and the use of that power can and will affect the physical and social environment in different ways. In the Challenger case the engineer was clearly accountable to the client/employer, and together they were responsible with, and to, many other groups. This included the government, the astronauts and their families and indeed any future astronauts. The nature of potential payloads extended responsibility further to wider society and the environment. This sense of shared responsibility is increasingly stressed in the light of environmental and global awareness. This wider environment cannot be the responsibility of any single agency. It demands shared responsibility.

This mutual responsibility for the effects of the project, negative or positive, leads to the practical question about what that responsibility requires in any situation. This may entail working through conflicting responsibilities and values, not least because there may be a multiplicity of 'stakeholders' (those who have a concern in or are affected by the project). In all this the wider society and the environment are not static entities but are continually changing. Hence, the engineer has to consider how to relate to them as part of his or her planning process. In the work on the Falklands Airport (Armstrong et al. 1999), for instance, the engineers were faced by the moral claim of the penguins, which far outnumbered the humans on the islands. Responding to their situation had to become part of the brief. Mining engineering is a particularly sensitive area of concern. Working across the globe, companies are faced with complex scenarios. These may include local and national governments, different tribal structures, all of which are affected by issues of ownership. Local environmental regulations and international concern for the environment are explored by nongovernmental organizations (NGOs). Nongovernmental organizations are interest and pressure groups, such as Christian Aid or Oxfam, that are concerned either about relieving global poverty and distress or about standing out against the behaviour of major corporations that may attempt to oppress local populations. The power of NGOs is such that they can mobilize public opinion and radically affect the public's perception of the companies involved. At the very least the engineering profession needs to be aware of them and should be prepared to consider and respond to their claims.

The professional body

So far we have argued that the engineer shares responsibility with the client or employer and that this relationship should demand that technical and ethical issues be reviewed and considered as part of any project. We have also argued that the work of the engineer is such that whatever he or she creates can affect the social and physical environment, sometimes drastically, usually for long periods of time. All of this suggests that the identity of the engineer is related to not just the individual operating in a vacuum but is actually worked out in relation to others. A critical 'other' in this respect is the professional body.

The message here is that the engineer is not alone. At one level the engineer contracts into the profession to protect his or her rights. This is a necessary negative freedom – a freedom from any attempt to undermine the engineer. Such a protective stance can of course move into what Ivan Illich (1977) argues is the dark side of the profession. Professional bodies can easily evolve into self-serving power blocks that, far from enabling client autonomy, encourage dependency upon the expert and thus discourage the responsibility of the client. The possibility of this presents an ongoing ethical challenge to the professional bodies.

However, the positive side of the professional body is its emphasis on standards that are both technical and relational, so that the professional engineer does not misuse his or her power. The stress from the professional body is not just on the development and maintenance of competence but on ethical standards expressed in their codes. The maintenance of such standards provides the basis for trust in the professional body and by extension in the professional. Hence, clients and employers can have confidence that where things go wrong the professional body will have an interest in correcting problems and in ensuring that all engineers learn from the failure.

The engineer in turn has a responsibility to the professional body and by extension to other professional engineers to uphold the standards shared by the profession. This is to ensure that trust is maintained and to ensure that the integrity of the engineering profession, with all the associated values of professional autonomy and responsibility, is maintained. In all this anything that the engineer does affects the profession. Equally, the professional's adherence to high standards can ensure trust from the public in spite of any particular problems with particular engineers.

The professional identity

The reasons for being ethical, in addition to self- and mutual interest, have begun to emerge from the complex environments and the relationships within which the professional engineer operates. From this a sense of shared responsibility can be identified. It could be said that this shared responsibility is reflected in the developing identity of the engineer. This identity has, of course, changed over time. Engineering was originally associated with the establishment and the power structures, political and spiritual, of the contemporary era. The engineer was required to produce for his masters the machines of war, cathedrals, tools and weapons. In his or her situation at those times there was little autonomy and no sense of responsibility towards any other – save the employer, the king or lord. It could be said that the engineer in that era was simply a technician who did as he or she was told. The consequences of nonperformance were dire. The engineers of Xerxes (Martin and Schinzinger 1989, 289), for example, were charged with the task of building a bridge across a waterway. A storm led to the loss of the bridge and the subsequent loss of the engineers' heads, which Xerxes felt was suitable encouragement for future engineers. With the development of science, scientific universities and material science and the greater demand for the technical skill of the engineer, the expertise and independence of the engineer was stressed, as it became increasingly difficult for clients to understand the work done on their behalf. With globalization, as we shall see in a later chapter, this has led to a complex environment within which the engineer operates. It is clear that the engineer's role has moved from purely technical to a broader professional one.

Airaksinen (1994), however, suggests that engineering is still not quite a profession. Indeed, he refers to it as a pseudo profession. His view of the 'proper' profession is one based on what are fundamentally moral values and the task is to work out those values explicitly. The doctor focuses on health and the distribution of health, the lawyer on justice and the teacher on education. The engineer, he argues, focuses on technical creation, and this is not a good in itself, an ethical value per se.

Against this, Davis (1998) suggests that engineering is not simply about technical creation, but is rather 'an instrument of material progress'. The engineering profession is committed to human progress, and this is explicitly articulated in the American Engineering Code that states that engineers 'use their knowledge and skill for the enhancement of human welfare' (Davis 1998, 15). This will be explored in more detail in Chapters 4 and 9.

Exercise

Along with two other students discuss how you see the identity of the engineer. Is concern for the development of general welfare part of that identity or is Florman right in saying that the engineer is concerned first and foremost with the clients' needs? What reasons would you give for your conclusion?

Conclusion

Self- and mutual interest are perfectly good reasons for the engineer to be ethical. Indeed, it could be said that they are necessary. An engineering professional or company will not be able to maintain or develop their practice without concern for the self. In this respect we can see that the engineer has the same needs as other stakeholders do. However, self-interest is not a sufficient reason to be ethical. Alongside self-interest is shared responsibility, with a stress on the shared responsibility for the social and physical environment (Armstrong et al. 1999, 31). We will return to a more detailed view of responsibilities in Chapter 4 when they are related to professional codes.

Such responsibility is not an optional extra, but arises directly from the identity and practice of the engineer. Engineering ethics is about how that identity and practice is worked out, with all the associated tensions and conflict. Chapter 2 explores in greater depth the meaning and foundations of such an ethics.

References

Airaksinen, T. (1994). Service and science in professional life. In *Ethics and the Professions* (R. Chadwick, ed.) pp. 1–13, Aldershot: Ashgate.

Armstrong, J., Dixon, J.R. and Robinson, S. (1999). *The Decision Makers: Ethics in Engineering*. London: Thomas Telford.

Bauman, Z. (1989). *Modernity and the Holocaust*. London: Polity.

Burleigh, M. (2000). *The Third Reich: A New History*. London: McMillan.

Davis, M. (1998). *Thinking Like an Engineer*. Oxford: Oxford University Press.

Florman, S. (1976). *The Existential Pleasures of the Engineering*. New York: St. Martins.

Harris, C., Pritchard, M. and Rabins, M. (1995). *Engineering Ethics; Concepts and Cases*. New York: Wadsworth.

Illich, I. (1977). *Disabling Professions*. London: Marion Boyars.

Koehn, D. (1994). *The Ground of Professional Ethics*. London: Routledge.

Lee, S. (2003). *Uneasy Ethics*. London: Pimiloco.

Martin, M.W. and Schinzinger, R. (1989). *Ethics in Engineering*. New York: McGraw Hill.

May, W. (1985). Adversarialism in America and in the professions. In *The End of Professionalism?* [Occasional Paper No.6.] pp. 5–19, Edinburgh: Centre for Theology and Public Issues, University of Edinburgh.

Reber, R. (2001). Virtual games inviting real ethical questions. In *Technology and Ethics* (P. Goujon and B. Heriard Dubreuil, eds) pp. 121–32, Leuven: Peeters.

Rees, L. (2005). *Auschwitz*. London: BBC.

Report of the Presidential Commission on the Space Shuttle Challenger Accident (1986). Washington DC: US Government Printing Office.

Robinson, S. (1992). *Serving Society: The Social Responsibility of Business*. Nottingham: Grove Ethics.

Robinson, S. (2002). Nestlé baby milk substitute and international marketing. In *Case Histories in Business Ethics* (C. Megone and S. Robinson, eds) pp. 141–58, London: Routledge.

Vardy, P. (1989). *Business Morality*. London: Marshall-Pickering.

2 *The foundations of ethics*

Introduction

For millennia we have had countless people standing in the marketplace of ethics telling us what they think should be the ethical foundation of our life journey doctrines, metaphysical theories (theories about the underlying nature of reality), ethical theories and the like.

In this chapter we will:

- Trace the developments of ethics from its early association with religion to autonomous ethics.
- Outline the postmodern era, when ethics seems to have become relativized.
- Trace alongside this the historical development of the engineer.
- Critically examine ethical theories.
- Note developments in ethics, not least the ethical systems based on virtues.
- Develop the virtues in relation to engineering and the professional skills.

Ethics

Exercise

Discuss with a fellow student:

1. What do you see as the foundation of your ethical practice?
2. What informs your understanding of right and wrong, good or bad behaviour?
3. Where does that understanding come from? Your family, your culture, your community, your religion, your school, your place of work, your university?
4. What difference do your different foundations make to each other?

For the engineer in the Middle Ages the foundation of ethics was not a problem. As we saw in the Chapter 2 the engineer developed his or her role in relation to the major contemporary power structures. The kings and princes in turn developed their role in relation to their key belief systems. Constantine (325) had realized that a shared belief system, in his case Christianity, would provide the best means of underpinning the unity of his Western empire. Kings had a divine right to rule, something harking back to the Jewish tradition of the Kings. It is precisely this tradition to which Handel's Coronation Anthem (1727) 'Zadok the Priest' alludes. Even today the coronation in England takes place in Westminster Abbey.

For the engineer in Europe of the Middle Ages, then, there was little doubt that God was on the side of the King, and there were no issues or ambiguities about professional ethics. You did as you were commanded and right was defined by the religious belief system, which was set out in the different scriptures. Islam sets out some ethical parameters, including advice on business ethics, which is strongly against usury (Hadith: Mishkat, 2799: Koran 2:276). The Judeo-Christian scriptures communicate the word of God through:

- Stories and history that illustrate key ethical concepts or attributes.
- Codes that sum up core ethical principles.

The Ten Commandments are a good example of a code, revealed by God to Moses, by which many people claim to live.

The Ten Commandments

- Worship only God
- Make no images of gods
- Do not blaspheme
- Keep the seventh day holy
- Respect your parents
- Do no commit murder
- Do not commit adultery
- Do not steal
- Do not give false testimony against you neighbour
- Do not covet your neighbour's possessions or relationships (Exodus 20:3–17)

More broadly, religions provide a framework of metaphysical meaning within which core values and motivations for being moral are set out. Metaphysics involves beliefs about the ultimate reality of the world. In the Judeo-Christian tradition, for example, there is the doctrine of the 'fall', the way in which humanity distanced itself from God. This points to a view of man as innately sinful, and suggests that humanity is not capable of acting ethically without the help of God. Another major doctrine is about the end of time and how God will, in some way, judge all. This provides a powerful motivation, with the options of heaven or hell a direct consequence of how we choose to behave in this life.

Different religions had different ethical stresses and there were, and are, different viewpoints about what ethics is within each religion. Nonetheless, ethical meaning was tied to religious beliefs, and there was broad agreement amongst religions about the core ethical attitude, summed up as respect for common humanity and the so-called 'golden rule'. Some of the versions of this include:

- *Christian*. 'Treat others as you would like them to treat you' (Luke 6, 31). 'Love your neighbour as yourself' (Matthew 22, 39).
- *Hindu*. 'Let not any man do unto another any act that he wisheth not done to himself by others, knowing it to be painful to himself' (Mahabharata, Shanti Parva).
- *Confucian*. 'Do not do unto others what you would not want them to do to you' (Analects, Book xii, 2).
- *Buddhist*. 'Hurt not others with that which pains yourself' (Udanavarga, v.18).
- *Jewish*. 'What is hateful to yourself do not do to your fellow man. This is the whole Torah' (Babylonian Talmud, Shabbath 31a).
- *Muslim*. 'No man is a true believer unless he desires for his brother what he desires for himself' (Hadith Muslim, imam 71–2).

Gradually, however, the association of ethics and religious belief began to erode.

- Metaphysics, especially in its supernatural form, could not be validated empirically.
- It is not easy to discern the voice of God in scriptures, which are a product of culture, expressed through laws, rules, songs and stories. The difficulty with religion and cultural context is neatly summed up in the following excerpt of an amusing open Internet letter.

A letter posted on the Internet to Dr Laura, a hard-line Christian broadcaster who argued that homosexuality is wrong, based on Leviticus 18.22. The writer 'accepted' Dr Laura's view on homosexuality, but wanted some clarification on some of the other Old Testament laws.

Dear Dr Laura,

- I would like to sell my daughter into slavery, as it suggests in Exodus 21:7. In this day and age, what do you think would be a fair price for her?
- Leviticus 25.44 states that I may buy slaves from the nations that are around us. A friend of mine claims that this applies to Mexicans, but not Canadians. Can you clarify? Why can't I own Canadians?
- I have a neighbour who insists on working on the Sabbath. Exodus 35.2 clearly states that he should be put to death. Am I morally obliged to kill him myself?
- My uncle has a farm. He violates Leviticus 19.19 by planting two different crops in the same field, as does his wife by wearing garments made of two different kinds of thread. He also tends to curse and blaspheme a lot. Is it really necessary that we go to the trouble of getting the whole town to stone them (Leviticus 24. 10–16). Couldn't we just burn them to death at a private family affair like we do with people who sleep with their in-laws (Leviticus 20.14)

I am confident you can help
Your devoted fan, Jim

None of this is to subvert religious ethics. Nonetheless, it shows that any religious command needs to be interpreted and critically tested, not accepted regardless of context or reason. Hence, ethics cannot be simplistically dependent on religion. Plato already pointed to this in the dialogue *Euthyphro*. In effect, he asks, is a thing good because God wills it or good because we agree that there is a rational basis for it. But how can something be good just because someone says it is, even God? On what can their judgement be based? The upshot is that ethical judgements have to have a reason, and this makes the basis for those judgements independent from God. Hence, the argument is that ethics is autonomous – free from religion.

In practice, so many religious prescriptions are based on reasoning. A good example is the Ten Commandments. Exodus describes them as coming from God. However, they can simply also be seen as pragmatic good sense. A relatively small group trying to make its way across inhospitable terrain of the desert would have needed a code of practice in order to maintain discipline and thus survive.

With the Enlightenment, a time when writers such as Kant (1724–1804) stressed the capacity of man to make rational decisions, the autonomy of ethics was further reinforced.

Engineers

Meanwhile, whilst ethics was becoming loosened from the ultimate reliance on religion, the discipline of engineering was developing. Firstly, across Europe there was a move away from engineers as royal employees to a group of experts who had skills that required training. This included the development, in the eighteenth century, of schools for engineering in Prague and France. It is in the nineteenth century that the full extent of the industrial focus of engineering developed, based in England, with the rise of the industrial revolution, not least through the development of steam engine technology that was applied to mining, railways, mills and so on. With these developments came the first UK engineering society, the Institution of Civil Engineers (1818). The engineer was closely associated with technological progress and a real pride in the capacity of nations to expand in this area.

The later part of nineteenth-century engineering stresses even more the identity of the engineer as a technical expert, with greater emphasis on technical and scientific training. Technological progress promoted the many different engineering specialities, thus diminishing the idea that engineering has one role or a single voice (Grelon 2001). Meaning and purpose were therefore focused on the role of the engineer as an expert, and also on the role of the engineer as an enabler of technological progress. Thomas Tredgold of the Institution of Civil Engineers was thus, in 1828, able to define civil engineering as follows:

> Civil Engineering is the art of directing the great sources of power in Nature for the use and convenience of man. . . . The most important object of Civil Engineering is to improve the means of production and of traffic in states, both for external and internal trade.
>
> (Davis 1998, 15)

In all this, engineering became associated with a very different kind of power, involving both modern democracy and the free market.

Ethical theories

Meanwhile, the study of ethics was focusing on moral theories as foundations of ethics. These were the rational foundations for ethics. The two most important were the utilitarian and deontological theories.

Utilitarian theory

This theory suggests that we find the meaning of ethics by looking at the consequences. In particular something is right if it maximizes the good, producing the greatest good for the greatest number (Mill 1806–73). The classic example of this theory is the scenario of the portly potholer who is stuck in the cave exit, with fifty of his colleagues behind him. If there is no alternative but to use explosives to blow open the hole then it is acceptable to sacrifice him for the good of the majority. The stress in all this is social good. Hence, if we look at abortion, for instance, we cannot say that this is wrong in itself. It can be right if it maximizes utility. In this case the argument would be that making abortion legal saves the lives of many pregnant women, who in the 1960s only had recourse to back street abortionists, and enables greater life choices for women as a whole.

Principles can be a rule of thumb, but cannot be absolute, and can reasonably be broken if they do not produce the most good. But what is the good? Mill's answer to that is happiness. All other goods are simply a means to that intrinsic good.

There are several different kinds of utilitarianism, including:

- Act utilitarianism, focusing on the act that is likely to produce the most good.
- Rule utilitarianism, focusing on sets of rules or codes that would produce the most good.

Utilitarianism is close to the mindset of the engineer who is looking to maximize the good in technological progress. Trying to base all ethical judgements on this utilitarian theory, however, is problematic. Firstly, Mill argues that happiness involves the higher, not the base pleasures. But who decides what is a higher pleasure? The very concept of the good that should be maximized is not clear. Secondly, the stress on everything as a means to happiness can easily subvert morality, with the end justifying any means. Can it be right to kill the populations of two whole cities in order to end the Second World War? Can it be right to accept or give bribes to keep a workforce employed? Can it be right to use torture in order to save lives? At the very least, such examples raise real conflicts about means. Thirdly, the stress on good for the majority can easily lead to the oppression of the minority. Just because something will benefit 50.5 per cent of the population does not make it good, and might in any case disadvantage the 49.5 per cent. Utilitarianism then is a useful tool but taken as a single moral theory it offers more questions than answers.

Deontological theory

The deontological approach to ethics argues that duties are the base of ethics rather than the consequences. Right actions, according to Emmanuel Kant (1964), are prescribed by duties – keep promises, be truthful, be fair, avoid inflicting suffering on others and reciprocate the kindness of others. Kantian ethics is thus about doing the right thing regardless of whether it makes one happy, quite the opposite of Mill's view. Kant also see duties specific to the self, such as do no harm to the self, develop character and skills.

What makes such prescriptions duties? Kant suggests that they:

- embody respect for persons,
- apply without qualification to all rational persons, and
- are universal principles.

What makes a person worthy of respect is the capacity to be rational and to develop the good that will enable the person to do their duty and to fulfil their purposes. This respect involves treating people as ends in themselves with their own purpose and capacity to fulfil it. This in turn means treating people not as a means to our ends. Coercion and manipulation of different kinds exhibit disrespect in these terms, such that the other is only useful for what they can do for you, with the aim to ensure that they do that.

This leads to certain key moral imperatives. Kant contrasts these, referred to as categorical imperatives, with nonmoral imperatives, which he refers to as hypothetical. Hypothetical imperatives are commands that are based on a hypothesis. 'If you want to get fit then exercise regularly' is one example. Categorical imperatives do not have such conditions. It is simply wrong to cheat or to break a promise. These are basic principles that are true without any reference to conditions.

Such principles should apply universally, and Kant argues that the most common principles pass this test. 'Promises should be kept', for example, applies in all situations. If we did not keep promises then the very meaning of the word would be questioned.

For Kant this points to a view of ethics that is based upon absolute principles, not because they are from the word of God but because they have this rational foundation.

Once again, however, it is difficult to see this as an exclusive foundation for ethics. Yes, it makes sense to stress the responsibility and duty of the person. However, firstly, absolute principles are very hard to pin down. The principle of promise keeping is justified by saying that if people do not keep promises then the meaning of promise keeping is eroded. But this is true only if promises are broken regularly. It is possible to say that, all things being equal, promises should be kept, the reasons being that they form the basis of a contract. However, it may be possible that a person has to break a promise because of some greater concern. For example, a person may have promised to maintain confidentiality only to realize that the person involved is a murderer who could kill again. Equally it could be that a person is not able to keep a promise. For example, a person may promise to support a particular project but lose the resources, personal or financial, to do this. In both these situations it could be argued that it is wrong to keep a promise. Behind all this it is very difficult to find any principle that does not have an exception to it.

Secondly, absolute principles have the danger that they discourage the responsibility of the individual for moral reasoning in context. This can lead to a lack of awareness of the situation, with principles applied uncritically.

Thirdly, the idea of reason as the basis for respecting another person excludes all human beings who are unable to reason intellectually, not least the severely learning disabled. The very word 'person' is as much a word that expresses value as it is a description. Hence, there is real danger of excluding some humans from that respect if it is at all conditional, i.e. if respect is based on some particular aspect or property of the person. The example of the Holocaust shows no respect for many groups of people outside the Third Reich and hence the inability to see them as people.

Intuitionism

Ranged against the two examples of rational foundations to ethics, Hume (1975) had little time for the place of reason as a foundation for ethics. We must rather look to the heart, the passions. Reason could provide the rational justification of means, but nothing can provide a rational justification of ends. The promptings of the heart are no excuse for ignorance, and reason has to guide our understanding of the world and the possibilities of that world, but intuition is very much at the base of what we determine to be good. Hume's position suggests that reason and emotion are critical parts of any foundation of ethics.

Questions

Which of these major ethical theories informs your ethical practice?
Which of them might best inform ethics in engineering, and why?

Postmodernity

With the twentieth century came the rise of postmodernity and a challenge to all attempts to find a foundation in ethics. Scholars argue fiercely about postmodernity and what it might mean (Connor 1989). Postmodern theory suggests new ways of understanding language and social constructs with the natural foundations of knowledge rejected, and no correspondences between language and reality. The human person is no longer autonomous but rather the effect of discourse and power systems (Baudrillard 1983). With no objective sense of reality each person has to create their own reality and underlying life meaning. Whether or not we agree with such theory, it is hard not to accept something of the postmodern *experience*, in which old certainties have broken down:

- A breakdown of any sense of objective knowledge and in particular adherence to the so-called 'grand narratives' of the twentieth century and before. These are the 'stories' that claim some universal truth and that influence whole generations (Lyotard 1979). For example, in England before the First World War (1914–8) there were the grand narratives of the Empire and of Christianity. The first of these saw the Empire as a force for good, both civilizing the world and enabling technological progress. Christianity, especially focused in the Church of England, was seen as the basis for most social meaning and support. Hence, each Church of England parish looked after the parishioners from birth, through marriage, to death. The First World War and ensuing economic and health crises destroyed those cosy views.
- A breakdown of patterns of behaviour and institutions such as marriage and the family, caused partly by increased wealth and mobility, and changes in cultural attitudes. This has led to massive increase in the divorce rate, an increase in cohabitation rather than marriage and the increasing acceptance, legally and culturally, of single sex partnerships
- A greater awareness of cultural and religious diversity within society, caused by increased migration and global awareness (Markham, 1994). This has led to a greater acceptance of ethical plurality, i.e. that ethical meaning is based in many different cultures and can lead to very different practices.
- A greater acceptance of a liberal view of ethics. Such a view argues that one can do anything as long as it is does not affect another adversely. This has partly developed as a result of the New Age, in the late twentieth century. The New Age involved many different groups that argued against morality being prescribed for them, by Church or by State. The individual person had every right to work out for themselves their value and belief systems (Perry 1992).

All of this has tended to unpick ethics from community and tradition, with shared foundations sliding into rights-based morality, and the lack of a shared ethical understanding. Ethical relativity means that there is no agreed approach to ethics. Each person or group will work out their own ethical meaning. The only thing that binds them together then becomes an agreement to tolerate the different views, unless they lead to the harm of another.

Exercise

1. Examine your university or business. Can you see in any overall shared ethical understanding?
2. What are the main values in the institution? Is there anything that sets them apart. How do they connect to policy and practice?

Rediscovering meaning

Many philosophers in the late twentieth century argued against ethical relativity and for a shared meaning in ethics. The key movements were virtue ethics, discourse ethics, global ethics and post-Holocaust ethics.

Virtue ethics

Alasdair McIntyre (1981) suggests that we must choose between Aristotle and Nietzsche. Aristotle locates ethics in an intelligible framework that makes sense of ethical dialogue. Nietzsche suggests that the old moral terminology no longer binds us; indeed, because it robbed us of the freedom to determine our own values, it *should* no longer bind us. McIntyre plumped for Aristotle. Ethical meaning, he suggests, is situated in a community of practice, and is communicated not through principles but through stories, and I would add all explicit and implicit expressions, including community ritual, that sum up the key virtues of the community. And it is virtues that are at the heart of ethical meaning. In Aristotelian terms virtues are dispositions for action that occupy the mean not the extreme. Courage, for instance, is neither cowardice nor foolhardiness. Underlying the virtues is the *telos*, purpose, which could involve well-being or happiness. One of Aristotle's key virtues (intellectual, not moral) is *phronesis* the capacity to reflect on the underlying good. In all this, the approach is to get the character right and good ethical practice follows from this. Unlike Kant, who saw the good decision as a matter of the will, not a matter of emotions and therefore of inclinations, Aristotle saw the ethically correct person acting out of the inclinations that the virtues gave him. Such virtues are learned through practice.

Virtue ethics is an important move forward in ethical theory, but has its problems. Firstly, it does not really get over the problem of how to handle ethical relativity. If each community is the basis for ethical meaning then there can be no sense of shared ethics beyond that community. Secondly, and connected, community-based ethics cannot inform any idea of justice, which by definition goes beyond boundaries. This is exemplified by universal human rights, the belief that certain rights apply to all regardless of their culture or community. Action that abuses those rights is seen as unjust. Thirdly, virtue ethics assumes a community that has a single voice and view of the good. In fact, all communities are polyvocal, with very different voices and perspectives. Ironically, this is particularly clear in religions. As Meeks (1993) notes of the Christian church, it developed, and continues to develop, its ethical meaning through disagreement and debate.

Discourse ethics

Awareness of this plurality led to a greater stress on the need for dialogue as essential for discovering ethical meaning. Habermas (1992) suggests that ethical meaning emerges from dialogue, enabling reflection on values and the discovery of shared norms. Getting the process right for such dialogue is thus of the highest importance, and Habermas suggests basic conditions for this. Benhabib (1992) goes further, noting that whilst the dialogue may reveal shared moral meaning, the conditions of dialogue themselves already embody moral meaning, not least of which is respect. This in turn requires attributes such as empathy (Benhabib 1992, 52).

Charles Taylor (1983) builds on all this. He places moral evaluation at the centre of human identity. Persons understand their identity in large measure by the 'strong evaluations they assert about what is good'. In other words, by actually asserting what you believe to be good you reveal something about your identity. This understanding in turn directs our lives. Taylor stresses the direct connection between morality and identity, between value and self-value/esteem. The self involves self-interpretation, and dialogue:

> My discovering of my identity does not mean that I work it out in isolation, but that I negotiate it through dialogue, partly overt, partly internal, with others. . . . My own identity crucially depends upon my dialogical relations with others'.
>
> (1983, 67)

This is all about the development of agency, responsibility and consciousness through dialogue. Taylor acknowledges the importance of community and tradition, and also the understanding of plurality in the self and in the community.

Global ethics

Global ethics builds on the sense of connectedness stressed by feminist ethics, and stresses our responsibility for the environment and human kind globally. This focuses on mutual responsibility for the environment and for different issues on a global scale, and a concern for structural global response to issues that dwarf the all too often individualistic focus of ethics (Kung 1991). It also reinforces that diversity in ethics, in this case cultural, has to be taken seriously. It cannot be assumed that a Euro-centric view of ethics can be dictated to a whole other culture, such as that of China. Respect demands that we listen to other cultures and that we understand what they have to say, and find ways of working with them.

Post-Holocaust ethics

This brings us right back to Willy Just, to engineering and gas trucks. In post-Holocaust ethics, writers such as Zygmunt Bauman (1993) reflect on the experience of the Holocaust. The Holocaust happened because of the way in which certain groups of people were excluded from humanity. Responsibility for 'the other' was, in these cases, denied. This denial was exacerbated by management techniques such as the division of labour, which further distanced any sense of responsibility. The basis of ethics, Bauman argues, has to be inclusive awareness and appreciation of the other. Hence ethics *begins* with taking responsibility for the other, and the rest is how that responsibility is worked out, with others. This forces us to think in terms of a shared responsibility and how that responsibility can be shared in practice. Too often responsibility is seen only in individual terms and quickly denied. A good example of this is the Herald of Free Enterprise, noted in Chapter 1. Because responsibility was interpreted in a narrow way they were able to live with the idea that superiors were responsible for the policy and practice, despite their awareness of the unfolding tragedy.

It is clear by now that no single view of ethical meaning and theory is satisfactory, but that the ethical meaning is found through using all of these:

- There is need for an irreducible ethical attitude that expresses concern, does not exclude and takes responsibility for ethical response and impels us to work out responsibility together. This can be summed up in the positive sense of the golden rule, but is more than a rule. It involves an awareness and appreciation of the other, including the social and physical environment.
- Ethical principles are useful and important concepts that sum up basic values. These can include respect, fairness, freedom, equality and so on. However, the exact meaning of these only comes alive in relation to each other and to the practical situation.
- Utilitarian calculus is important, such that full account of consequences can be taken.
- There is need for dialogue that will engage value and attitude differences within and between different communities. There are few situations that do not have many different values expressed by different stakeholders. Not only will this enable greater awareness of the ethical identity of others involved in a situation,

it will also clarify and strengthen one's own ethical identity, and identify how values can be embodied and also challenged in practice.

- Virtues are essential in ethics, enabling reflection on values, and appropriate response. It is hard to put ethics into practice without virtues. Virtues, however, are always instrumental in that they serve a particular end (telos). Hence, they require both a fundamental ethical attitude and principles to fully give them meaning.

All this provides a much more manageable view of ethics for the engineer, combining both awareness of others and their needs, awareness of one's own values and the capacity to critique and challenge one's own and one's community's values and the values of others. Yes, the intercultural breadth of modern society does involve a great diversity in ethical viewpoints. Nonetheless, with a core ethical attitude, a reflective method based around dialogue that values but can also challenge traditional ethical views, attention to consequences and the development of related skills and virtues, there is both the tool and the character to navigate the ethical waters of the twenty-first century.

In the rest of this chapter we will begin to develop virtues and ethical principles in relation to engineering. In Chapters 3 and 4 we will look at the method of ethics and how this might relate to the practice of project work, the professional principles and also the codes that try to sum up ethical practice.

Virtues and the engineer

Virtues can easily be seen as something quite separate from the skills and attributes that lie at the base of training in technological competence. The work of the engineer Richard Carter strongly suggests that in fact the virtues are integrated into the professional skills.

A useful starting point for reflection on these connections is Carter's (1985) taxonomy (Table 2.1).

From the perspective of engineering, Carter argues for the need to develop personal qualities alongside skills. The way in which these qualities are characterized could be debated. Nonetheless, the divisions help in reflecting on how the ethics of character relates to the professional's task. Firstly, the attitudes and values section is primarily about valuing or respecting the other, from ideas to persons, to the environment. Secondly, in the midst of personality characteristics is what is often seen as a critical ethical virtue, integrity. Solomon (1992, 168) suggests that integrity is not one but a complex of virtues 'working together to form a coherent character, and identifiable and trustworthy personality'. Thirdly, Carter refers to spiritual (not exclusively religious) qualities. This involves awareness of the self

Table 2.1: A summary of a taxonomy of objectives for professional education (Carter 1985)

Personal qualities	**Mental characteristics**	**Attitudes and values**	**Personality characteristics**	**Spiritual qualities**
	Openness	Things	Integrity	Awareness
	Agility	Self	Initiative	Appreciation
	Imagination	People	Industry	Response
	Creativity	Groups	Emotional resilience	
		Ideas		
Skills	**Mental skills**	**Information skills**	**Action skills**	**Social skills**
	Organization	Acquisition	Manual	Cooperation
	Analysis	Recording	Organizing	Leadership
	Evaluation	Remembering	Decision-making	Negotiation and persuasion
	Synthesis	Communication	Problem solving	Interviewing
Knowledge	**Factual knowledge**		**Experiential knowledge**	
	Facts		Experience	
	Procedures		Internalization	
	Principles		Generalization	
	Structures		Abstraction	
	Concepts			

and the other, and the capacity to respond to the other, be that a person, group, environment or transcendent other. At the base of spiritual qualities is something about empathy, the capacity to reach out and identify with the other. At the same time, such qualities also invite us at least to reflect upon the connection between values and belief systems (in the broadest sense of systems of practice and belief that give core meaning to people).

We can now consider how these and other virtues relate to the core responsibilities of the engineer. These can be summed up in terms of the responsibilities of the engineer to the client/employer, the public/humanity and to the professional body.

Client/employer

The client/employer relationship requires trust. This is developed through the efficient and competent use of technical knowledge and skills, through confidentiality, and through communication of the truth, which enables the client/employer to make a fully informed decision.

This demands several virtues including:

- Veracity: This is not simply truthfulness or the avoidance of lying but the capacity to communicate the truth such that the client can make an autonomous and informed decision.
- Perseverance: This in necessary both for the acquiring of technical competence and for the development of solutions.
- Loyalty/fidelity: This involves the capacity to develop trust in the client, such that the engineer continues to work in his best interests. If virtues are about the mean between extremes then loyalty to the client is between the extremes of an inability to sustain a relationship and unthinking adherence to client/employer. Hence, it is not clear that loyalty demands that the professional engineer should do whatever the client wants if this involves illegal or unethical behaviour or ends.
- Alongside such virtues are a number of intellectual and personal qualities that relate to relational skills and the skills of organization, analysis, evaluation, synthesis, communication and problem solving. These include creativity, imagination, flexibility, initiative and industry.

Public/humanity

Behind this responsibility is something that links to the spiritual qualities of Carter: awareness of the other. This is partly straightforward awareness of the context, and of associated stakeholders, in which the engineer operates. This involves concern for the health and safety of the other – avoidance of harm – and also concern for the positive good of the other.

May (1987) suggests a virtue of public spiritedness, something that orientates the professional to the common good. This ties in with the stress in secondary and higher education in citizenship and awareness of the person's part within society.

Professional body

This relationship demands:

- Capacity to cooperate with others and to be aware of how the professional body relates to the common good.
- Respect for the worth of what the professional and his or her colleagues produce. This is partly a proper view of the value and pleasure found in the work of engineering. In Chapter 9 we will look more closely at proper professional pride.

- Humility. This might seem the oddest of virtues for any professional whose work is based in technical excellence. However, this virtue involves a realistic assessment and acceptance of limitations as well as strengths. The mean here is between false modesty and boasting. It is an important virtue in responsibility negotiation and working in teams.

Overall virtues

There are certain virtues that enable the professional engineer to work through moral reasoning and to handle the conflicts that may emerge between the different relationships and responsibilities noted above. These include:

- Temperance: This involves the capacity for moderate behaviour, balance and self-control. It is important in enabling a measured response to any crisis and for effective judgement. As such it supports self-reliance and helps to focus on responsibility.
- Justice: This involves the capacity for fairness. This is partially about equal regard for different groups in any professional context and partly about responding to need.
- Courage: This involves the mean between cowardice and foolhardiness. This virtue could be central to the many aspects of the professional's work, from the courage to whistleblow to the courage to stand out against a client's preferred option.
- Hope: This is one of the so-called theological virtues, which also include faith and love (St Paul's Epistle to the Romans, chapter 13). Hope in this context means the capacity to envision a creative and significant future, something central to the creative work of the engineer.

Central virtues

Core virtues that can be drawn from Carter's taxonomy are respect or care, integrity, practical wisdom and empathy.

Respect

Respect is the basic ethical attitude. It involves inclusivity and unconditionality. May (1987) notes that these are basic to the idea of the professional, summed up in the idea of the promissory covenant. The UK medical profession as a whole, for instance, aims to be always available and not to demand conditions, such as ability to pay, for treatment. Nonetheless, as May also notes there is need for a specific contract to ensure appropriate and equitable treatment. Respect involves a commitment to the other.

Integrity

Integrity is a complex of virtues, involving several aspects:

- *Integration* of the different parts of the person: emotional, psychological and intellectual. This leads to holistic thinking and an awareness of the self alongside awareness and appreciation of external data.
- *Consistency* of character and operation between value and practice, past, present and future, and in different situations and contexts. The behaviours will not necessarily be the same in each situation, but will be consistent with the ethical identity of the person.
- Taking *responsibility* for values and practice. Without accepting responsibility for ethical values and for response neither the individual nor the profession can develop a genuine moral identity or agency.
- *Independence* is a key virtue. It ensures distance such that the professional can stand apart from competing interests and more effectively focus on the core purpose. This enables professional autonomy.

Absolute integrity is impossible to attain. Hence, an important virtue is humility. Equally important therefore is the capacity to reflect, to evaluate practice, to be able to cope with criticism and to alter practice appropriately. This capacity to learn means that integrity should not be seen as simply maintaining values and ethical practice, come what may, but as involving the reflective process, such that values can be tested in the light of practice and either appropriately maintained or developed. Failure to maintain integrity gives rise to conflicts of interest and values that can lead to termination of employment (Sims and Brinkmann 2003).

Practical wisdom

This is very close to Aristotle's intellectual virtue of phronesis, the capacity for rational deliberation that enables the wise person to reflect on his or her conception of the good and to connect this to practice. Aristotle sees this not as a moral virtue but as an intellectual virtue. This virtue is often the one most tested when targets have to be met. In the Challenger case the invitation to look at things from a manager's perspective was inviting the engineers to consider a different underlying purpose to their actions, involving stakeholder demands and tight timetables. The purpose underlying the engineer's role was quite different, safety.

Empathy

Empathy is closely connected to the virtue of benevolence and enables the professional to identify with the other. If wisdom is an intellectual virtue then empathy is an affective (to do with feelings) virtue. It is the capacity to hear and understand underlying feelings. It does not mean total identification (sympathy) but rather enables an appropriate distance between the self and the other. Such a distance is necessary if the other is to be understood, and if the professional is to operate impartially and effectively (Robinson 2001, 56). Empathy is important in handling different relationships. It enables the professional to see why different views might be held so strongly, and thus enables him or her to respond appropriately to them.

Empathy and practical wisdom connect the concern of ethics with professional competence. Both enable appreciation and care of the other and both support imagination, creativity and openness, key to the creative process and in making effective decisions. In turn, qualities such as imagination can affect awareness of the other and thus the development of empathy. Such virtues and capacities enable holistic thinking and practice, and thus effective engagement with the ethical issues at the heart of the professions and business.

Respect, integrity and empathy are important in developing character and identity, built around what Mustakova-Possardt (2004, 245) refers to as critical moral consciousness. She sums this up as involving four dimensions: a moral sense of identity, a sense of responsibility and agency, a deep sense of relatedness at all levels of living and a sense of life meaning or purpose. These core elements of ethics resonate strongly with Yorke and Knight's (2004) four components of what is necessary for employability:

- *Understanding*. This is intentionally differentiated from knowledge, signifying a deeper awareness of data and its contextual meaning.
- *Skills*. This refers to skills in context and practice and therefore implies the capacity to use skills appropriately.
- *Efficacy beliefs, self theories and personal qualities*. These influence how the person will perform in work. There is some evidence, for instance, to point to malleable, as distinct from fixed, beliefs about the self being connected to a capacity to see tasks as learning opportunities rather than as opportunities to demonstrate competence (Yorke and Knight 2004, 5). This in turn influences commitment to learning goals and the capacity to learn.
- *Metacogntion*. This involves self-awareness and the capacity to learn through reflective practice.

Such virtues can be learned only through practice and in particular in a community of practice that has a shared sense of purpose. It is precisely in such a community that any professional body seeks to be, developing a culture that itself could be said to embody virtues such as integrity. Importantly, focus on the virtues is a reminder that ethics involves more than just rational reflection. Ethical practice involves an awareness of underlying emotions, both connected to the purpose of the profession and in dealing with any ethical issues. Hence, the community of practice in many countries involves more than simply accreditation. A good example of this is the ceremony of the Iron Ring, worn by Canadian Engineers. This is given to graduating engineers as part of a ritual that is reminiscent of the doctor's Hypocratic oath. The ritual and the wearing of the ring are means of raising and maintaining the engineer's consciousness of the high standards, responsibilities and professional pride embodied in their community of practice. Legend has it that the original rings were made from the steel beam of the Quebec Bridge that collapsed in 1916, simultaneously reminding the engineer of the need for humility and responsibility.

Conclusion

Ethics is not about simply applying principles or calculating possible consequences. It involves the whole person in reflecting on underlying values and in responding to personal and professional relationships. As such it is not an add-on to work but is a core part of the culture and decision-making of the professional engineer, and hence the identity and character of the engineer and the engineering professional bodies. The underlying values are embodied in the decision-making and culture of the individual professional and the decision-making of professional bodies and engineering businesses – the communities of practice.

Ethical practice as part of such a community involves several things:

- It is essentially a learning experience, not the application of inflexible principles, and thus relates directly to reflective practice.
- It has a strong social aspect to it. The idea of integrity can be used for a person or an organization (Armstrong et al. 1999). Also, no one can embody all the virtues. They are more likely expressed and distributed within a group or team. Ethics itself is seen as increasingly collaborative, i.e. not simply about the individual working out a moral dilemma but about how responsibility is shared.
- The stress on virtues, empathy and so on is essentially one that looks to the development of ethical autonomy, not ethical prescription. The person is responsible for the reflection of, commitment to and development of whatever values they hold, and for any response.

- The development of such agency and moral identity is always in relation to social structures and the very different values systems that are part of anyone's personal or professional history. Development of moral autonomy therefore demands working through how one relates to the different value systems. To do this demands an awareness of these value systems, the capacity to enter into a dialogue with them and the capacity to both appreciate and challenge them. In terms of moral development, this means moving away from an unquestioning acceptance of moral norms to a critical appreciation of how they relate to other moral perspectives and to particular situations.

The focus on character, deliberation, practice, identity and agency enables the professional to identify core values that give meaning and purpose to his or her work, ranging from professional autonomy to inclusive care. At the same time, it enables the professional to handle creatively the very diverse values embodied in different stakeholders within a particular community and beyond. Handling such plurality is at the heart of global ethics.

In all this, competence and ethical behaviour are not opposites but are both necessary for the professional. It is an ethical imperative to maintain standards and thus competence, just as it is an imperative to treat others with respect. Some might be surprised at the connection of moral virtues to the kind of professional skills that have been central to professional training for decades, such as the skills of problem solving, communication, analysis, creativity and imagination. However, the relationship is close. As we noted in the Challenger case, transparency was key to good ethics, and transparency is not possible without careful communication. Creativity, imagination and the capacity to collaborate are directly useful in finding ways of responding with integrity to the demands of stakeholders. Equally, virtues such as empathy and integrity enable a better awareness of the needs and values of those involved in any project and thus enhance competence. Important to all of this is how the professional makes decisions and how these virtues are a part of that decision-making. It is to the method of ethical decision-making that we will turn in Chapter 3.

Rudyard Kipling

Kipling (1865–1936) was a poet and author who identified strongly with the role of the engineer, such that he was invited in 1922 to create the ritual of the Canadian engineers' ring referred to above. Kipling is most famous for the Jungle Book stories and the poem 'If . . . '

His poem, the Sons of Martha, focuses on the role of engineers. Martha and Mary refer to the New Testament story of Mary and Martha with Jesus. Martha was the one who did all the work and Mary was the one who sat at the feet of Jesus, just listening. In this poem Kipling sees the engineer as the son of Martha. They work through and develop everyday technology, they struggle with the environment, they have to keep coming back to work through technical problems, they seek to serve. Somewhere behind this is the suggestion that the engineer is not always appreciated.

The Sons of Martha

The Sons of Mary seldom bother, for they have inherited the good part.
But the Sons of Martha favour their Mother of the careful soul and troubled heart.
And because she lost her temper once and because she was rude to the Lord, her Guest,
Her Sons must wait upon Mary's Sons, a world without end, reprieve or rest.

It is their care in all the ages to take the buffet and cushion the shock.
It is their care that the gear engages; it is their care that the switches lock.
It is their care that the wheels run truly; it is their care to embark and entrain,
Tally, transport, and deliver duly the Sons of Mary by land and main.

They say to mountains, 'Be ye removed'. They say to the lesser floods, 'Be dry'.
Under their rods are the rocks reproved – they are not afraid of that which is high.
Then do the hilltops 'shake to the summit – then is the bed of the deep laid bare,
That the Sons of Mary may overcome it, pleasantly sleeping and unaware.

They finger death at their gloves' end where they piece and repiece the living wires.
He rears against the gates they tend: they feed him hungry behind their fires.
Early at dawn, 'ere men see clear, they stumble into his terrible stall,
And hail him forth like a haltered steer, and goad and turn him till evenfall.

To these from birth is Belief forbidden; from these till death is Relief afar.
They are concerned with matter hidden – under the earthline their altars are;
The secret fountains to follow up, waters withdrawn to restore to the mouth,
And gather the floods as in a cup, and pour them again at a city drought.

They do not preach that their God will rouse them a little before the nuts work loose.
They do not teach that His Pity allows them to leave their work when they damn-well choose.
As in the thronged and the lighted ways, so in the dark and the desert they stand.
Wary and watchful all their days that their brethren's days may be long in the land.

Raise ye the stone or cleave the wood to make a path more fair or flat:
Lo, it is black already with blood some Son of Martha spilled for that:
Not as a ladder from earth to Heaven, not as a witness to any creed,
But simple service simply given to his own kind in their common need.

And the Sons of Mary smile and are blessed – they know the angels are on their side.
They know in them is the Grace confessed, and for them are the Mercies multiplied.
They sit at the Feet – they hear the Word – they see how truly the Promise Runs:
They have cast their burden upon the Lord, and – the Lord He lays it on Martha's Sons

(Rudyard Kipling, 1907. http://bellsouthpwp.net/s/b/sbahrd/kip4.html)

Exercise

1. Kipling's view of the engineer was coloured by how he saw engineering practised by the army in the British Empire. As such there are many assumptions about values that differ from today. How much of what he writes is true for the engineer of today? If you were asked to sum up the place and purpose of the engineer in today's world how would you put it?
2. Kipling's greatest poem 'If . . . ' sums up some core virtues. Read the poem and list the virtues referred to. How many of these are relevant to the engineering professional?

References

Armstrong, J., Dixon, J.R. and Robinson, S.J. (1999). *The Decision Makers; Ethics for Engineers.* London: Thomas Telford.

Baudrillard, J. (1983). *Simulations.* New York: Semiotext.

Bauman, Z. (1993). *Postmodern Ethics.* Oxford: Blackwell.

Benhabib, S. (1992). *Situating the Self.* London: Polity.

Carter, R. (1985). *A taxonomy of objectives for professional education.* Studies in Higher Education, **10**(2), 135–49.

Connor, S. (1989). *The Post Modern Culture.* Oxford: Blackwell.

Davis, M. (1998). *Thinking Like an Engineer.* Oxford: Oxford University Press.

Grelon, A. (2001). Emergence and growth of the engineering profession in Europe in the 19th and early 20th centuries. In *Technology and Ethics* (P. Goujon and B. Heriard Dubreuil, eds) pp. 75–100, Leuven: Peeters.

Habermas, J. (1992). *Moral Consciousness and Communicative Action*. London: Polity.
Hume, D. (1975). *Enquiries Concerning Human Understanding and Concerning the Principles of Moral*. Oxford: Oxford University Press.
Kant, I. (1964). *Groundwork of the Metaphyics of Morals* (J. Paton, trans.). New York: Harper and Row.
Kung, H. (1991). *Global Responsibility*. London: SCM.
Lyotard, J-F. (1979). *The Postmodern Condition*. Manchester: Manchester University Press.
McIntyre, A. (1981). *After Virtue*. London: Duckworth.
Markham, I. (1994). *Plurality and Christian Ethics*. Cambridge: Cambridge University Press.
May, W. (1987). Code and covenant or philanthropy? In *On Moral Medicine* (S. Lammers and A. Verhey, eds) pp. 83–96, Grand Rapids: Eerdmans.
Meeks, W. (1993). *The Origin of Christian Morality*. New Haven: Yale University Press.
Mustakova-Possardt, E. (2004). Education for critical moral consciousness. *Journal of Moral Education*, **33**, September, 245–70.
Perry, M. (1992). *Gods Within*. London: SPCK.
Robinson, S. (1992). *Serving Society: The Social Responsibility of Business*. Nottingham: Grove Ethics.
Robinson, S. (2001). *Agape Moral Meaning and Pastoral Counselling*. Cardiff: Aureus.
Sims, R.R. and Brinkmann, J. (2003). Enron ethics (Or culture matters more than codes). *Journal of Business Ethics*, July, **45**, no. 3, 243–56.
Solomon, R. (1992). *Ethics and Excellence*. Oxford: Oxford University Press.
Taylor, C. (1983). *Human Agency and Language*. Cambridge: Cambridge University Press.
Taylor, C. (1989). *Sources of the Self*. Cambridge: Cambridge University Press.
Yorke, M. and Knight, P. (2004). *Embedding Employability in the Curriculum*. York: LTSN.

3 *The practice of ethics*

Introduction

In Chapter 2 we looked at the foundations of ethics and began to develop the idea of virtues and engineering. We ended with the argument that the core foundation of ethics was the acceptance of shared responsibility. But if responsibility is shared then there has to be some process for working out just what that involves. There needs to be a practical framework for ethical decision-making. In this chapter we will develop such a framework, by:

- Looking at the work of the reflective practice school.
- Developing examples of ethical decision-making methods.
- Summing up professional ethics principles embodied in the method.
- Noting how ethical reflection runs throughout an engineering project.

These methods will form the basis of how one can respond to an ethical dilemma, and also of how ethical awareness can be a proactive part of planning, thus avoiding, as far as possible, dilemmas. Ethics in this context simply becomes part of the everyday operation and culture of the engineer.

Developing an ethical framework is important for several reasons.

- It provides a means of systematic ethical reflection, and thus a tool for both the student and the professional to practise.
- It enables the person to develop ethical autonomy, to take responsibility for ethical decision-making. Here it is not important for the person to accept a particular framework but rather to develop, and test through dialogue, his or her own.
- It enables the development of holistic thinking in which ethical framework (as process) and ethical content are not seen as separate. If the framework enables autonomy and the development of responsibility, then it is embodying core ethical values.

Such a framework is best related directly to any general decision-making process, so that ethics is located in everyday business or professional reflection. Many people when faced by the request to reflect on their method, especially if the person making the request is an 'expert' in ethics, feel uncertain. 'I really can't think that I have a method' is one typical response. But, of course, in practice we all have a method for dealing with issues of value and ethics in our life. We are simply not always aware of it. Hence, in order to develop an ethical framework it is important to begin articulating what our present practice is.

Exercise

Reflect on what is your everyday method of ethical decision-making. Share this with a fellow student and discuss any differences in your approaches.

Reflective practice

The process of reflective practice can be best seen as a recurring learning cycle involving several phases, summed up by Kolb's (1984) learning circle.

- *Articulation* of narrative in response to experience.
- *Reflection* on and *testing* of meaning.
- *Development* of meaning.
- *Response,* leading to new experience.

Donald Schoen (1983) develops this further. He famously argued against a simplistic view of professional development. Much professional education of the old school assumed that there was a standard body of knowledge that would be learned in professional schools. This knowledge would be applied to a range of issues. The old view of the medical profession, for instance, was one of paternalistic doctors applying their expertise without consulting the patient. The problem with this approach is that:

- It can encourage a view of professional practice that is largely about achieving targets.
- It can lead to the attempt to apply knowledge in an unreflective way, hence to impose expertise.

Schoen noted, through observation of a range of different professions, that in practice there was a response that led not to an imposition of knowledge, but rather to a 'reflective conversation with the situation'.

What emerged was a process like this:

- The analysis of the situation in order to work out what the problem might be and what issues are involved.
- 'Appreciative' or value systems that help to find significant meaning in the situation.
- Overarching theories that might provide further meaning.
- An understanding of the professional's own role in the situation, both its limits and its opportunities.
- The ability to learn from 'talkback'. This involves reflective conversation about the situation.
- The professional would also treat clients as reflective practitioners.

In a sense Schoen is simply developing the idea of what it is to be a professional, building on the work of Ivan Illich. Illich (1977) questioned the role of the professional in society, not least in the way that society has become dependent upon professionals such as doctors. Such dependence was negative, partly because it led to the individual not taking responsibility and partly because, through lack of awareness, it led to personal and public disasters. Perhaps just as important was the simple idea that the technical skill of the professional could not be exercised without taking into account the relational context.

Gibbs (1988) provides another simple framework that takes account of the emotions as well as ideas in reflection:

- *Description*: what happened?
- *Feelings*: how did you feel about the situation?
- *Evaluation*: what was good or bad about the situation?
- *Analysis*: what was good of bad about the situation?
- *Conclusion*: what else could you have done?
- *Action plan*: if this arose again what would you do?

Schoen and Gibbs do not fully draw out the ethical dimensions of or values behind reflective practice, but it is possible to do so. The underlying values of reflective practice include:

- The person taking responsibility for his or her own ideas and values, and how they relate to practice.
- Responsiveness to the situation, enabling dialogue with the client and stakeholders.
- Awareness of the professional's role and limitations.
- Respecting the autonomy of the client.
- The importance of continued learning.

Each of these can be seen as an ethical value. Awareness of limitations, for instance, can be seen in terms of humility. It is also of great practical importance for any professional. If a surgeon, for instance, operates unaware of a physical limitation, or of limitations in experience, this could put lives at risk.

In one sense reflective practice is primary. The ethical dimension of this comes through strongly in the principles, values and belief systems that are brought to the understanding of the situation and to the evaluation of the different options. It is at these points that ethical reasoning becomes important, in knowing what the core values are and how to relate these to possible consequences. Ethical reasoning also involves critically examining the underlying moral values to see what role they play in any decision and how they can be justified. The basic moral imperative 'it is wrong to lie', for instance, is accepted by most people most of the time, but could it be right to lie in some circumstances? A response to that would demand close examination of the situation, reflection on consequences and reflection on the rights of those involved.

It is to a framework for working out what is right in practice that we now turn.

Ethical reflection

To explore how the values and beliefs are handled we will work through a simple case study.

Case 3.1

A Western civil engineering firm with links to a Colombian private university was asked by the Colombian government to lead a project with the aim of meeting the technological needs of tribes in the remote rainforest. After a long and difficult journey the engineers, with a number of students, reached the village and agreed that the most urgent priority was an adequate supply of water. The nearest water supply was a large river, about half a mile from the village, and some 15 m lower than it.

The crops on which the villagers depended required a regular and large supply of water, which was not always easily available during the drier parts of the year, – though the river did continue to flow during these times. The journey to the river was not pleasant and involved wading through a swamp The swamp area had been increasingly used as a latrine by the tribe. The swamp area was also infested with mosquitoes and was the cause of disease. Nonetheless, this area is in an environment that includes major flora and fauna, including two rare animals.

The village, of course, had no electricity supply, and there was only a rudimentary technology. Housing conditions, in huts made of timber and willows, were very poor. Illness and deformity were common, and the villagers had virtually no contact with the outside world.

The culture, however, was an attractive one. Honesty was a key virtue, and the villagers were accustomed to work hard to keep together body and soul. The corporate spirit of the villagers was very strong, supported by a lively belief system and frequent community rituals.

It seemed unlikely that anyone would attempt to support or further assist the village once the project was completed.

(This case was originally noted by John Cowan.)

How would you make decisions about this situation? Share your method with a colleague and work out a method of ethical decision-making that you are both comfortable with. Explain why that method is fit for its purpose.

This is an actual case. Because it involved students, many students in the first and second year are able to identify with it rapidly. The use of this case with students has led to a general framework of ethical reflection involving data gathering, value management, responsibility management, reflecting on options and planning and implementing.

1. *Data gathering*. The starting place of ethics is reflection on the situation. For the engineers in this case this meant looking at the stakeholders who were involved in the situation and what their interests and needs were. The immediate problem for them was the lack of clarity about who was the client. The government had commissioned the action and the university was funding the student involvement, but the tribe was the party that would directly benefit from the work. It was clear that clarification was needed. It was also clear that

the data collection stage is not purely technical. It is about finding out the perspectives and needs of the clients and also the other groups or areas affected in the situation. For the government this would be a major publicity opportunity. They were being criticized for their lack of concern for the rainforest and its inhabitants. They wanted this to demonstrate the care of the government. For the civil engineering company and the university, getting this right could have affected any future developments in which they hoped to be involved in Colombia. At the centre of it all was a tribe whose views were not clear. How soon would Western, so-called developed, culture affect them? How can one speak of enabling the tribe to develop an informed decision in this situation? Most of these questions can be looked at in the context of the professional developing a relationship of trust with the tribe. This would crucially involve working with government representatives who would have to stay focused on the immediate problem, with a clear statement about the limitation of resources. Surrounding all this was the rich environment of the rainforest, which was under threat. The environment itself could be seen as a stakeholder having an interest in the engineering response. Core to all data gathering is clarity about the expectations of stakeholders. Hence, gathering data can depend upon building up relationships with those involved in the situation. Whose responsibility that is will be a matter for negotiation (see below).

2. *Value management*. The second stage of ethical decision-making is the identification and working through of the values involved. In this case there are clearly several different kinds of values that the engineer needs to be aware of:

   - *Personal ethical values*. An engineer may have particular concerns for the environment or for the health and well-being of groups in the undeveloped world. Some students who have responded to this case have been so concerned about the plight of the tribe that they argued that the engineers should look into developing a proper healthcare system for the tribe. The question of how personal values relate to any case then emerges. In some cases they may be worthy, but not part of the role of the engineer. In other cases the personal values may supersede professional values. Hence, some engineers might find it difficult to work in contexts where there is the potential for abusing the rights of stakeholders. Either way it is important to identify any personal values and how they might relate to the situation.
   - *Professional ethical values*. Professional moral values can be seen as fundamental and procedural principles:

     (a) *Fundamental principles*. Fundamental values of the profession include *nonmaleficence* and *beneficence* (Beauchamp and Childress 1989). These are very broad principles, meaning above all do no harm, and also do good. Interpreting these in the situation then becomes important.

In this case this meant that the engineers had to ensure that they did not adversely affect the lives of the tribe, in terms of either their culture or their health or environment. This applied to the environment as a whole, which was under threat in that area. Again, this would have to be worked through with the other interested parties. Another basic principle suggested by Beauchamp and Childress is *respect for autonomy* (self-governance). This can be seen in two ways. Firstly, in a negative sense it involves not interfering with the freedom of the clients to decide for themselves. Secondly, in a positive sense it refers to enabling the clients to make their own decision. Respecting the tribe's autonomy then involves respecting any decision they come to. The tribe, for instance, might politely decline any offer of help, in which case the engineers would have to inform their other 'clients', the university and the government, that the project, for all its importance, could not come off. Positive respect for autonomy would involve a careful discussion with the tribal leaders to work through things that the engineers had to offer. This would include the limitations of any offer, such as insufficient resources to maintain any sophisticated technology over a long period. Some engineers in responding to this case felt that the very fact of the Western group being there altered the dynamics of the situation and would inevitably affect the culture and environment for the worse. A final basic value is of *justice*, such that the engineer treats all stakeholders fairly.

(b) *Procedural principles.* These are the important values that inform the professional's engagement with the situation and are really instruments of the basic principles. Confidentiality is essential for respecting the autonomy of the client. Such confidentiality needs to be formally established. How would that work through the case above, where so many stakeholders and possible clients are involved? Informed consent or decision-making is also essential for positive respect. In addition, Beauchamp and Childress argue for cost–benefit analysis and risk–benefit analysis as ways of assessing consequences in the light of the basic principles.

- *Public values.* This involves two things. The first view of public values is the values of the different groups. In this case this might lead to conflicts in values between the engineer, the government and the university. A second view of public values is really values that underlie public life. In this case such values include community, freedom and health. Student groups who have discussed this case have noted the strong tension between the Western view of health, largely about absence of disease, and the view of health espoused by the tribe, which stresses on health in terms of personal and cultural coherence.

Summary of professional principles

Fundamental principles

- Respect for the autonomy of the client
- Justice
- Beneficence
- Nonmaleficence

Procedural principles

- Confidentiality
- Informed consent or decision-making
- Cost–benefit analysis
- Risk–benefit analysis

David Seedhouse (1998, 128 ff.) argues against such principles. This reminds us that any view of ethics is contested. He suggests that the fundamental principles are not clear enough. There are, for instance, many different views of justice. However, Seedhouse misses the point of such principles. They must be quite general, to give some idea of values that can be shared across different areas. At the same time, the clearer meaning of the principle emerges as one engages with the situation. The procedural principles also help to give meaning. What is clear is that principles cannot be applied in an unthinking way, i.e. without reference to the situation.

- *Value conflict*. Values need to be clearly articulated, at which point there may be congruence and agreement between values or value conflict. Value conflict may emerge in terms of a dilemma. A dilemma is where there is a choice only between two equally bad alternatives, or where there are two equally important values that need to be fulfilled. This is distinguished from an ethical problem, where there seems to be a solution. The first key response to any apparent conflict is to clarify exactly what is being said. In one sense this should be part of the data gathering. Some engineering practitioners now make a written summary of what they think the client actually wants. This can be explicitly confirmed through the client's signature. Often this shows that the ethical problem or dilemma is actually inferred, not real. Sometimes, however, clarification confirms the dilemma. The classic example of a dilemma is whistleblowing (see Case 3.2). The other classic value conflict situation is *conflict of interest*. Conflicts of interest may occur in different ways but are usually seen in terms of a personal interest or activity of the engineer that could influence, or might appear to influence, the engineer's capacity to act in the best interest of the client and profession. Another way of viewing this is an interest that might affect the engineer's professional autonomy. A classic example of this was in the Challenger case with conflict between the safety-centred engineering interests and the interest of the firm to keep their contract with the US government.

Some apparently conflicting values are in fact capable of being held together, whereas others may preclude mutual operation. Yet other values, such as autonomy and 'best interest', need to be constantly monitored. Where there is dialogue about different values, this will lead to exploration of the justification of the value, and thus the nature of any belief or value foundation. Reflection on why values are held will reveal a range of reasons, some rational and some deeply held beliefs with emotional content (Cowan 2005). It is at this point that it becomes important to be aware of the underlying strength of feeling, which may be fuelled by psychological, cultural or religious experiences. This reflection can take place in any situation and can concern local community issues; multicultural issues in and beyond the community and multinational issues, for instance, in the developing world. Dealing with a complex ethical issue may well necessitate the development of skills in conflict resolution that enable different parties to work through feelings as well as ideas (Harvard Business Review 2000). There are, therefore, important links between personal and moral development and the acquisition of interpersonal management skills.

In the case above some of the core values needed to be held together, such as autonomy of the client and concern for community, and welfare of the environment and the tribe alongside the limitations of resource.

Responsibility analysis and negotiation

Research on family social policy and families has noted the importance of negotiating responsibilities in developing and maintaining ethical meaning. Finch and Mason (1993) concluded that a majority of families did not work from principles or any predetermined value base. Rather, they negotiated responsibility and through this developed ethical reputation and identity. A similar approach can apply to this phase of decision-making. Firstly, it involves identifying the stakeholders in any situation. Secondly, there is an analysis of the stakeholders in terms of power and responsibility. This enables a full appreciation of constraints and resources in the situation. Thirdly, responsibility can be negotiated. This does not simply look to the development of goods for all stakeholders. Rather, it accepts the premise of mutual responsibility and enables its embodiment. Hence, it enables a maximization of resources through collaboration and the development of an ethical identity that can include all those involved. Negotiation skills are therefore important.

In the case above, responsibility had to be clarified between the engineers, the university, the government and the tribe itself. This in turn would help to clarify issues about resources.

Responding

In making a response that takes account of the ethical issues the various possible options need to be reviewed in the light of the different values, bringing together value and principles, consequences, responsibilities and dialogue amongst the stakeholders. This also requires awareness of the constraints in any situation and the resources maximized by collaboration.

In this case there were three options: to do nothing; to develop an intermediate technological solution that would fully involve the tribe and to develop a more sophisticated solution that would require high maintenance. The actual engineers opted, with the tribe, for the middle option, in the form of primitive piping and an Archimedian screw, because it would mean:

- Minimal interference with the values of the tribe.
- The tribe retaining its independence.
- The tribe retaining its community.
- Positive effects on the health of the tribe.
- No major imbalance with other tribes in the area.
- Minimal interference with the environment.

Questions

How far were the professional values fulfilled in this solution? How far were the interests and values of the other stakeholders fulfilled?

Of course, the deadlines of work mean that embodiment of values cannot be 'perfect', and the fitting of values to practice will be a matter of ongoing reflection.

Indeed, given such constraints any ethical response may involve choosing the best of several difficult options. In medical ethics this is exemplified in the case of conjoined twins, where at best one of them would die (Lee 2003). Nonetheless, the use of a framework that is understood and accepted by colleagues can enable rapid critical reflection for the immediate situation. Clearly, such a framework may not be used in detail for every situation. Moreover, the experienced reflective practitioner may well use the framework without consciously working through it in a serial way (Eraut 1994). However, it provides a means of making ethical sense of any situation. It embodies and reinforces a view of ethics that is dialogic, participative, collaborative and transformative, and therefore directly relevant to employability.

Case 3.2

A civil engineer sees colleagues from his own firm finishing off a side road into a nearby estate without constructing a proper foundation. Should he tell his superiors?

The value conflict that emerges in this simple case of whistleblowing is between loyalty to the colleagues, loyalty to the firm, loyalty to the profession and concern for the future well-being of the people on the estate. Clarification of such values helps to set out what is involved. Well-being may not seem to involve major issues. The road will simply degrade over a period of time and be renewed. But its poor condition could lead to an accident.

Then come three possibly different views of loyalty. Loyalty to the profession involves ensuring that standards are maintained. Loyalty to colleagues involves a desire not to get them into trouble. Loyalty to the firm involves a commitment not to blacken its name. At this point the meaning of these different loyalties has to be clarified, not least because firms can use lack of loyalty as a major criticism of the employee, often leading to termination of employment or negative reactions from colleagues. However, the term loyalty does not of itself mean unthinking adherence to company practice or policy. On the contrary, it is more like a virtue of the middle, lying between unthinking obedience at the one extreme and lack of commitment at the other. Such a reflective commitment will have limits. Such limits cannot always be set out in any contract and may have to be explored in a particular situation. In light of this, anyone concerned about whistleblowing will look to find ways of responding to the problem in ways that respect those involved in the case. In this particular case the engineer could begin with a stance of respecting his colleagues, letting them know what he had seen and invite them to take responsibility for rectifying the issue. Swift corrective action might be possible. If this is unsuccessful then he can tell his colleagues that he will have to raise it with the firm within a stipulated time, again giving them the space to take responsibility. If, eventually, the whistle had to be blown to the firm or beyond, then the engineer would need to find ways of ensuring that the interests of his family were properly addressed. However, it would be in the interest of the firm to be aware of any lowering of professional standards. Hence, there is an increasing acceptance of the need for working cultures that enable positive whistleblowing (further explored in Chapter 6). The development of such cultures is a good example of negotiating responsibilities such that firms, for instance, enable the anonymous communication of information. This particular case shows something of the need to clarify values and to work out responsibilities. It also shows that a great deal depends on how any ethical response is framed. Where whistleblowing is adversarial it can lead to an escalation of negative responses, with bad effects for both the firm and the whistleblower. A more positive approach (Borrie and Dehn 2002) encourages the potential whistleblower to work carefully through the firm before the last resort of going outside. Ethical responses then are not simply about applying ethical values – they also need careful attention as to how the ethical values are embodied in the response, with a particular concern to work creatively through possible conflict situations.

Summing up so far

An ethical framework such as what we have suggested is:

- A space for reflection without having to be held rigidly to stages.
- A learning process. The professional does not apply ethical value unthinkingly, but discovers the meaning of values and principles in reflective practice.
- Essentially dialogic, listening to the other ethical value and interests, clarifying their meaning for the different stakeholders. Hence, such a framework is essentially social, not simply confined to the individual.
- Collaborative and practice centred. The different stakeholders become involved and the response is creative, often transforming the situation.

Another ethical decision-making method is set out by Michael Davis (1998) in Table 3.1.

Table 3.1: Seven-step guide to ethical decision-making

1. State the problem. For example, 'there is something about this decision that makes me uncomfortable' or 'do I have a conflict of interest?'
2. Check facts. Many problems disappear upon closer examination of the situation, whilst others change radically.
3. Identify relevant factors. For example, persons involved, laws, professional code and other practical constraints.
4. Develop list of options. Be imaginative, try to avoid 'yes' or 'no' but list things like whom to go to and what to say.
5. Test options. Use tests such as the following:
   - Harm test: does this option do less harm than alternatives do?
   - Publicity test: would I want my choice of this option published in the newspaper?
   - Defensibility test: could I defend this option before a committee of peers?
   - Reversibility test: would I still consider the choice of this option good if I were adversely affected by it?
   - Colleague test: what would my colleagues say when I describe my problem and suggest this option as my solution?
   - Professional test: what might the profession's governing body or ethics committee say about this option?
   - Organization test: what does the company's ethics officer or legal counsel say about this?
6. Make a choice based on steps 1–5.
7. Review steps 1–6. What would you do to make it less likely that you would have to make such a decision again?
   - Are there any precautions you can you take as individual?
   - Is there any way to have more support next time?
   - Is there any way to change the organization (e.g. suggest policy change at the next departmental meeting)?

Project audit

This chapter has so far investigated the practical framework for ethical decision-making in a general context. It will now be used to provide an overview of ethical reflection in the specific context of a project, from inception to completion.

Owner's initial idea of project. Determination of goals and performance criteria. Project manager appointed.
Project funding review and initiation. Preparation of first broad preliminary cost estimates.
Development and refinement of owner's outline specification: location, technical specification, performance specification, programme. Outline costs of construction and operation. Start of development of 'brief package'. Appointment of in-house or external project team and advisers/consultants.
Feasibility study. Conceptual design. Preliminary cost estimates and programme.
Preliminary design works. Development of cost estimates and programme.
Detail design works. Refinement of programme and project cost estimate.
Choice of procurement system. Development of tender package.
Definitive estimate by project team.
Tender Period. Development of contractor's estimate of project costs based on tender documents. Contractor's review of programme.
Submission of tenders. Tender review. Contractor's costs and programme against project team's estimates.
Client's decision to proceed. Award of contract.
Preconstruction period. Mobilization. Preliminary orders placed.
Construction phase. Payments to contractor. Costs and programme monitored.
Completion, commissioning, handover, start-up.
Operational life. Maintenance regime and repairs by operational staff.
Eventual operational shutdown. Decommissioning. Demolition

Figure 3.1: Typical life cycle of project.

Diagram 3.1 shows the key stages of the overall life of a project from the 'idea' to the end of the life of the project. The diagram can be used as the basis for a review of each of the critical audits carried out during the project.

To audit is 'to conduct a systematic review'. During the life of any project there are many such audits. Many are familiar. There will almost certainly be almost continuous cost audits, as the project is developed in the planning stage, then monitored in construction and operation. Safety and value as well as quality and risk may be audited. This section proposes that there should also be an ethical audit for the project.

Many parties are involved directly and indirectly in the various stages of the project. These can include:

- The owner and the owner's project team, his or her financial team and the operational and maintenance teams.
- The owner's project team may be in-house or may be supported by advisers, consultants, designers and other specialists who will provide assistance during the planning and design of the project.
- A separate in-house team may also be involved to advise on the operation of the project during the planning stage. This team will run and maintain the project after completion.
- A constructor or contractor's team will be given the task of building and commissioning the project. These may include various specialist suppliers and subcontractors. For the whole life of the project there may even be a demolition contractor to decommission and take down the project construction at the end of its use.
- Wider stakeholders will also be involved and will form a significant body of opinion. Stakeholders include employees, suppliers, users, neighbours – in fact, all those parts of society in general who might be affected, beneficially or adversely, by the project. This very process of identifying and naming of the stakeholder group is a very positive action in itself. It recognizes the interests of people who might be disadvantaged by the construction and operation of the project and effectively gives them a voice in the project.

It can be appreciated that the number of individuals and groups affected by, or involved in, the project can be large, and that their periods of active involvement may vary over the life of the project. All stakeholders have different perceptions of the project and many concerned or affected by the project may have different ethical agendas. All involved, from the chief executive officer or owner of the project, the financial officers and the staff down to the stakeholders themselves will come to the project with their own interests and values. In the context of the project, however, the construction teams and advisers may also be bound by group or company codes of ethics. Ideally, these company codes should encourage transparency and disclosure and should invoke standards of honesty, integrity, fairness, accountability, consideration for others, reliability and citizenship – all in pursuit of excellence in the chosen field of endeavour.

It can be seen that individual, personal and group interests are at work. Any attempt to establish relationship networks and communication systems, both formal and informal, involving all concerned will be difficult but will enhance the development of the project and increase the chances of its satisfactory conclusion. The identification of stakeholders and formation of these networks and communication systems are therefore the starting point for an ethical audit process.

The ethical audit of a major construction project is fully explored in Armstrong et al. (1999, chapter 6). One of the authors, James Armstrong, describes his personal involvement in the Mount Pleasant Airport Project in the Falklands Islands. He describes the process of attempting to identify stakeholders and their agendas at the outset of the project – rather than be confronted by them and having to try to accommodate them during the later and more complicated and costly phases of development of the project by removing potential areas of conflict.

The range of ethical problems involved is identified as the motives and scope of the project are developed and a common agreement of all concerned is achieved. This, however, is seen to be potentially very difficult or even impossible in the light of some projects that may typically have political, social, economic, physical, environmental and implementational aspects that may concern stakeholders. The problems of conflicting rights – rather than right and wrong – further complicate many situations.

It is appreciated that a project will also be viewed from both a current and a retrospective viewpoint and the attitudes and opinions of owners, advisers and stakeholders may well vary with time or prevailing conditions. Armstrong et al.'s ethical audit recognizes that 'engineering projects inevitably impact on the physical, economic and social environments that contain them. Environments include all factors outside the project that may affect or be affected by the project, not only green issues. These impacts may be beneficial or harmful' (p. 105).

What then is involved in the detail of a project ethical audit? How is the ethical behaviour described so far in this book put into practice? The problem addressed is the potential conflict arising in each of the different stages of the project, the interested stakeholder parties and those actively involved in the project. The process attempts to reduce potential conflicts by careful proactive planning and investigation and recognition of the interests and responsibilities of all concerned.

Diagram 3.2 shows the following five aspects of project realization:

- The interests and roles of the several decision-makers
- The brief
- The context of realization – political, social, economic physical, technical, legal and environmental
- The design
- The implementation

1. The decision-makers
Notes on parties involved – decision-makers advisers, other parties
Interdisciplinary team working
Authority – extrinsic, intrinsic
Motives, aim, objects
Conflicts – divergent or convergent negotiations

2. The brief
Wants
Needs
Constraints

3. The context
Societal context
Physical context
Constraints

4. The project design
Long-term effects – designer responsibilities
Reversibility – designer/owner responsibilities
Sustainability – all responsible
Maintenance – owner responsibilities
Environmental impact assessment

5. Implementation
Team structure and procedures
Control systems monitoring quality assurance. CDM regulations
Construction issues
Programme
Labour and material policies
Equipment

Figure 3.2: The project ethical audit. Source: From Armstrong et al., (1999, 93).

Each of these involves decisions with ethical implications and, as the project proceeds a larger group of people with more widely varying interests emerge.

The decision-makers. This takes into account the attitudes of the client and the team assembled by the client, consultants, suppliers and contractors, with many different backgrounds and skills. The ethical audit will help to identify the roles of all different parties and define their responsibilities within the project.

The brief. The project involves a great many constraints – technical, legal, financial, environmental, organizational and ethical – and the brief will define the priorities and objectives of the project, taking these constraints into account. The brief is developed through an interactive dialogue between the client and advisers. As trust grows between the parties not only are the technical elements developed, but the social, environmental, financial and ethical aspects of the project are also explored, agreed and recorded. These records form the full working picture of the whole of the project and can then be used to inform all involved, form cost and time models and become the basis of tenders for supply, construction, operation and maintenance.

The context. Engineering projects impact on the physical, economic and social environments that contain them. Engineers are responsible for the effective use and management of resources and a series of analyses that review the project within the context of a complex and changing environment. Environmental aspects are increasingly more important, rigorous and demanding.

The project is planned, constructed and operated in a system of formal regulations, institutional rules and personal ethics. The government exercises wide-ranging powers over all aspects of design, construction and operation, and over the broad range of statutory regulations, legislation covering health and safety, labour, security, traffic and transport and many other areas.

Engineers have the task of differentiating between personal views and professional duties and making objective analyses of projects. These analyses may in certain situations result in conclusions that may be justified in engineering terms but may offend the opinions and moral beliefs of others. All of the engineer's opinions must be based on sound professional knowledge and must be justified as appropriate to clients or the public. The engineer must retain the integrity that, in the event of his or her opinion being disregarded, they will follow their responsibility and duty and withdraw from the project, retaining the right to make their concerns known to the appropriate authorities.

The social environment of any project is not simply confined to the immediate social context. The materials chosen to develop the project may themselves have a history, in terms of how they have been manufactured or whether they have denied human rights to those involved. Such groups can be seen as stakeholders, significantly affected by the project. As we shall see in the Chapter 8, responding to their needs may take many different forms.

The project design. The design of the project is a development of the briefing process wherein the designer and the client work together to specify and define a project that meets the needs of the client and takes into account all the other external and internal project parameters. The balance between conflicting demands such as construction and operating costs, materials, deliverables, the requirements of different team members and stakeholders and the environmental consequences of decisions must all be considered and accommodated.

Implementation. As the project moves into the stage of implementation there are two areas of possible conflict.

1. Internal relations between the members or groups of the project team.
2. External relationships with the general public.

Internal relationships are usually well structured by this stage of the project, which is then running on the basis of formal contractual arrangements and operating procedures that are familiar to most project team members.

However, there will inevitably be stresses resulting from the pressures of maintaining time and cost control and the need for fast decisions. Clear organizational systems, levels of responsibilities and well-organized and maintained channels of communication all help to reduce the possibility of conflict.

The choice of contract itself can be a key issue. The traditional confrontational nature of construction industry contracts is slowly being superseded by attitudes of partnering and cooperation. At site levels, similar communication channels must be established and maintained so that good relationships between resident engineers, supervisory staff and site workforce are developed.

External relationships may become more problematic at this stage in the project. Wise project teams ensure that a good level of open consultation takes place with local interested parties during the planning stage, but there may be other members of the public who may not have been aware of these consultations – or even the project itself! These parties may only become aware of the scope of the project and its effect on their lives when physical work begins on site and they see the effects of factors such as a labour force and site traffic on local environment and facilities. This can be especially problematic in rural situations overseas where a large construction project can draw would-be workers, supplies and supporters to the general areas of the site – often creating difficulties for local limited infrastructure.

It is important for the project team to take all efforts to reduce the impact of the construction phase. Talks with local groups, liaison groups and committees and even 'hot line' communication systems have been carried out in the past for these purposes. If a sense of involvement in the project rather than a sense of alienation can be engendered in the local community then better public relationships result. As many engineering projects – particularly civil engineering projects – are intended to improve the quality of life of the community, this involvement may not seem such an insurmountable task.

The ethical audit procedure can now be seen as an entity, standing alongside the other more traditional audits of quality, time, cost and safety. The starting point is a clear definition of the 'plan' and a meaningful comparison with the facts of the actual situation – continuing throughout the life of the project. Armstrong et al. suggest the creation of an ethical file or database that is prepared and continues to be regularly updated throughout the stages of the project. The file represents a continuous appraisal of the situation, which can be used as a guide by the project team to learn about possible outcomes resulting from alternative lines of development. It is suggested that time is set aside specifically for consideration of ethical issues during the normal project meetings and that specific ethical review meetings are organized that:

- Review the present situation.
- Review the ethical implications of future activities.

These systems will clarify, define and identify possible sources of conflict throughout the life of the project and develop trust between parties involved. If this trust and genuine interest and respect for the concerns of external and internal parties involved in the project can be maintained, then the project is much more likely to be completed on time and to cost and with the approval of those affected by it.

Project exercise

In a group, use the Internet to investigate the case of the Bhopal Union Carbide Plant (1984). Place yourself in the position of the local board as they were setting up the plant. Detail your plans for this project in the light of the project ethical audit above.

Conclusions

In this chapter we have looked at the 'how' of ethics. Once again it becomes clear that this method is not wholly distinct from the everyday reflective practice and decision-making of the professional. Indeed, it would be odd if it were that different. Ethics would hardly make a difference in practice if the decision-making were that different. The ethical dimension is around the underlying values and principles of the engineer and the different parties involved, the values embodied in the consequences and the ethical identity of the engineer. The question of identity leads us to a consideration of professional codes.

References

Armstrong, J., Dixon, R. and Robinson, S. (1999). *The Decision Makers: Ethics for Engineers*. London: Telford.

Beauchamp, T. and Childress, J. (1989). *Principles of Biomedical Ethics*. Oxford: Oxford University Press.

Borrie, G. and Dehn, G. (2002). Whistleblowing: the new perspective. In *Case Histories in Business Ethics* (C. Megone and S. Robinson, eds) pp. 96–105, London: Routledge.

Cowan, J. (2005). The atrophy of the affect. In *Values in Higher Education* (S. Robinson and C. Katulushi, eds) pp. 155–77, Cardiff: Aureus.
Davis, M. (1998). *Thinking Like an Engineer*. Oxford: Oxford University Press.
Eraut, M. (1994). *Developing Professional Knowledge and Competence*. London: Falmer.
Finch, J. and Mason, J. (1993). *Negotiating Family Responsibilities*. London: Routledge.
Gibbs, G. (1988). *Learning by Doing: A Guide to Teaching and Learning Methods*. London: F.E.U.
Harvard Business Review (2000). *Negotiation and Conflict Resolution*. Cambridge, Massachusetts: Harvard Business School.
Illich, I. (1977). *Disabling Professions*. London: Marion Boyars.
Kolb, D. (1984). *Experiential Learning*. Englewood Cliffs, NJ: Prentice Hall.
Lee, S. (2003). *Uneasy Ethics*. London: Pimlico.
Schoen, D. (1983). *The Reflective Practitioner*. New York: Basic Books.
Seedhouse, D. (1998). *Ethics: The Heart of Health Care*. London: Wiley.

4 *Ethical codes*

What's the difference between mechanical engineers and civil engineers? Mechanical engineers build weapons

Civil engineers build targets

Anon

[W]e must remember that good laws, if they are not obeyed, do not constitute good government. Hence there are two parts of good government: one is the actual obedience of citizens to the laws, the other part is the goodness of the laws which they obey

Aristotle, Politics

Introduction

This chapter will investigate ethical and professional ethical codes. It will review:

- The purpose of codes, deciding at whom they are primarily aimed.
- Elements that are essential in a workable practical ethical code.
- The concept of the 'profession' and the 'professional code'.

As each profession and organization seem to have developed its own separate and individual code we will enquire why is this the case, and if it is possible to have only one single professional code for use by all? A brief methodology will be included to guide those wishing to write their own code.

A selection of existing engineering codes from the United Kingdom and overseas is appended to the chapter and these will be examined to highlight the issues raised above.

Background

In recent years there has been an upsurge of interest in the ethical behaviour of both business and the professions. Customers, clients and employees have begun to search for, and specifically try to work with, those organizations that have clearly set out their basic ethical ground rules.

This is especially evident in the case of financial institutions and companies with whom the members of the general public invest. Clear ethical statements are now published to attract ethical investors who try to ensure that their funds will be used in projects and organizations with acceptable and clearly stated ethical goals.

Increasingly, organizations are being judged not simply by their profit records but by their behaviour and their success in adhering to their stated ethical rules. They are held to account by the public, investors, clients and employees when they go astray.

What are the purposes of codes?

Codes have been developed by organizations and professional institutions and used in two main ways.

1. As guides to their members in the areas of ethical standards and standards of behaviour where they act as references in day-to-day decision-making.
2. They also act as statements of the organization's contract with society, setting out the ways that institutions and their members will behave in their everyday dealings with clients, employees, suppliers, communities and the general public.

The resulting codes have been refined and developed through use to become documents that aspire to enhance the ethics of all members of the organization and the organization itself. They actively focus the efforts of the organization in the selection and furtherance of their goals whilst recognizing their duties and responsibilities to the clients and communities they serve.

The codes therefore:

1. Define accepted standards of behaviour for the group.
2. Promote high standards of practice.
3. Provide benchmarks by which members can measure and develop their personal standards.
4. Define the ethical aspirations and identity of the group both internally and in relation to the public and communities around them.
5. Exhibit a level of maturity to the outside world.

Both the organization and the society benefit from the adoption of such codes. They develop an enhanced sense of community amongst the members and encourage a growth in the public approval and stature of organization for which they are written.

Professions and codes

It is to be noted that some of the organizations and codes mentioned so far in this chapter are described as 'professional codes'. It has already been suggested that engineering is a profession and later chapters will investigate the status of 'business' in a similar context.

In a profession such as engineering, whose works may both influence and strongly affect sections of the general public and even whole populations, it is especially important that a comprehensive code is developed and maintained to guide the members and reassure all of those affected by engineering projects. History has shown us that many engineering failures had their basis in ethical conflicts or in the containment of ethical issues.

Examples of seemingly high-level corruption in the construction industry worldwide are regularly reported. These are clearly ethical failures, but there are other relatively small-scale examples – failures in the work of apparently good, successful, hardworking professionals – which we may not think of in similar ethical terms at first reading.

For example, a reported study by the Swiss Federal Institute of Technology of 800 structural failure cases showed 504 persons killed, 592 persons injured and many millions of pounds accruing in damages (http://www.matscieng.sunysb.edu/disaster).

The causes of these failures were classified as:

Insufficient knowledge: 36 per cent
Underestimation of external factors: 16 per cent
Ignorance, carelessness, negligence: 14 per cent
Mistakes/errors: 13 per cent
Insufficient/inadequate control: 9 per cent
Imprecise definition of responsibilities: 1 per cent
Quality issues: 1 per cent
Other factors: 3 per cent

The majority of these listed items involve elements of human failures or design process failures but these same items can also be considered to involve failures of engineering ethics: engineers, acting as 'professionals', who have failed in their implicit responsibilities to their employers, their profession and to the general public by not using their skills and knowledge to the standards set by their organizations or professional institutions.

These broad responsibilities to employers and to society are not new; they were recognized by engineers in the early stages of the development of engineering as a profession. History shows that leading engineers throughout the world encouraged the inclusion of ethical standards in their professional charters. They recognized their roles and responsibilities in the industrial development of the societies in which they operated. These early professional groups include the Smeatonian Society of Civil Engineers founded in 1771 in the United Kingdom, the Institution of Civil Engineers founded in 1818 in the United Kingdom, the Boston Society of Civil Engineers founded in 1848 in the United States and many others. These forward-thinking groups recognized the responsibilities of the engineer to the societies on whose behalf they worked and publicly made ethical statements describing their intentions in carrying out their works.

These engineers recognized their responsibilities as part of their roles of being 'professional' and it is worthwhile at this stage in our review to formalize our own understanding of the terms 'profession/professional' so that we can fully understand its importance and its effect on the scope of any relevant ethical codes we encounter.

Professions

We live in a world of professional footballers, professional soldiers and doctors, lawyers, engineers and nurses – amongst many others – who describe themselves as professionals. What does this mean? Who are the true professionals? How is this status achieved? To whom and to what do true professionals owe responsibility?

A profession requires extensive training and the mastery of specialized knowledge. It often has a professional association or institution, a code of practice and a means for licensing. Examples are law, medicine, finance, the military, nursing, the clergy and engineering. The term 'professional' is often used for the acceptance of payment for an activity, in contrast to an amateur. However, the term 'professional' is also used to refer to practitioners who have a strong sense of service beyond their immediate client group (May 1985). Hence the *Oxford Shorter Dictionary* uses the term 'vocation' to define the term profession: 'A vocation, a calling, one requiring advanced knowledge or training in some branch of learning or science.'

To be a member of a profession is therefore to:

- Have specialized knowledge and skill.
- Have power – the power of knowledge and the capacity to affect society.
- Have autonomy of practice. This varies according to employment context.

- Have a monopoly or near monopoly of a particular skill.
- Have undergone an extensive period of training that includes not simply skills, but a strong intellectual element.
- To be a member of a professional body that is responsible for regulating standards, protecting rights of practice and ensuring proper training.

It is said that the professional engineer, in common with professional practitioners of all disciplines, needs six qualities:

1. *Integrity, openness and honesty*, both with themselves and with others.
2. *Independence*, to be free of secondary interests with other parties.
3. *Impartiality*, to be free of bias and unbalanced interests.
4. *Responsibility*, the recognition and acceptance of personal commitment.
5. *Competence*, a thorough knowledge of the work they undertake to do.
6. *Discretion*, care with communications, trustworthiness.

These professional qualities, linked with the personal virtues already discussed, enable the professional to maintain principles and fulfil responsibilities, and to do this in a context that is often complex and unclear. The professional may work in a discipline with specialized knowledge that his or her employer may not have or understand. He or she therefore works outside the control of the client and is trusted with the client's interests. The professional's qualities are therefore at the core of professionalism.

Integrity

This involves the discovery and communication of the truth. It is not, however, simply truthfulness or the avoidance of lying, but the capacity to communicate the truth in such a way as to enable the client and others to make informed decisions. Honesty and integrity are essential for the development of trust.

Integrity leads to a concern for the whole situation in decision-making, including an awareness of the professional's own attitudes, standards and value systems. It leads to consistency of character and operation in different situations and contexts. A good test of this is not just consistent operation in different contexts but also how the individual operates when there is nobody around, testing consistency and truth against personal attributes.

Integrity ensures that the professional does not accept 'moral distance' or the majority view of any particular group, but is true to his or her own objective understanding of the situation.

Integrity also involves the capacity to learn, as noted in Chapter 2.

Independence

Independence is not so much a virtue as a state. The true professional should be independent of pressure from any interested group. To be involved with either the employer or an action group in such a way that professional judgement is clouded is to lose professional autonomy. To achieve this, the professional must understand the situation and the key players in that situation. Independence enables the professional not to be drawn into the concerns of any particular group. It presupposes the willingness and ability to take responsibility for decisions.

By virtue of his or her special knowledge the professional has an intrinsic authority. He or she also has an extrinsic authority vested in him or her by virtue of the position held. It is important to recognize that such 'authority' leads many people to accept directions given. Their personal responsibilities may be denied and transferred to the authority figure. Professionals must respect and encourage others to accept their own responsibilities.

> Professionals have to have autonomy. They cannot be controlled, supervised or directed by the client. Decisions have to be entrusted to their knowledge and judgement. But it is the foundation of their autonomy, and indeed its rationale, that they see themselves as 'affected' with the client's interest.
>
> (Peter F. Drucker)

Impartiality

Impartiality enables the professional both to fulfil his or her contracts with the client and to treat all parties equally. There is a clear need to avoid concerns for self-interest, for concern for self-advancement and the pressures of management roles, because priorities for the achievement of programme and cost targets may cloud professional judgement.

Responsibility

Responsibility involves the realistic assessment of skills and capacities and the acceptance of their possibilities and limitations. This is a core virtue in enabling the professional to acknowledge the responsibility of others, to accept personal responsibility and to work as a team – an essential part of modern engineering projects.

Problems may arise if professionals develop an inflated sense of responsibility, seeing themselves as responsible for more than they can, or ought, to be involved with. Realism is critical for ensuring the best contribution from all members of the team.

From this we can see that the engineer owes responsibility to:

- The general public/society
- The users of the project
- The peer group of professional engineers and institutions
- Clients and employers

Engineers are responsible for:

- Their own actions.
- Duties accepted and implicit in their work and institutions.
- Legal requirements of the nations in which they practise.
- Legal obligations imposed by contract.
- The greater moral consequences of their actions, and of those for whom they are responsible.

On this basis many members of the construction industry, including architects, engineers, and quantity surveyors, are now described as 'professionals' and boundaries between their roles have become less clear. Consultants have a more commercial outlook, with marketing and fee-bidding strategies. Contractors are leading teams of 'professionals' in management contracts and in design-and-build agreements. Roles are further combined and interwoven by the addition of financial and legal specialists in private finance initiative (PFI) projects – finance, design, build, operate and hand over projects. All these have a considerable influence on concepts of professional ethics and responsibilities.

But this interweaving of roles requires an even more conscious understanding of the responsibilities of professionals. As professionals, engineers are still responsible for making decisions that others, without their skills and training, cannot make. The need for integrity, independence, impartiality and responsibility are even greater. It is important to avoid a crude view of professionalism that results in a concentration on the detail of the technical brief, ignoring social factors and the wider implications of one's decisions. The responsibility for one's role in collaborative work within and beyond the profession is of prime importance.

Competence

Perseverance is necessary for the acquisition of technical competence and for its application in solving technical problems. This involves the capacity to strive for and maintain competence in professional practice. Competence is essential if the professional is to fulfil his or her responsibilities to society. The ethical requirement is in the acknowledgement of the need to be competent, rather than in the competence itself. Skills can be misused.

An essential element of professional practice is the capacity to reflect upon, to evaluate and to accept different levels of competence, and to ensure that the necessary skill is deployed, even if it has to be procured elsewhere.

Discretion

In the course of projects the professional will become aware of many aspects of the affairs of clients, contractors or other interested parties, and it is important that they maintain their essential trust in him or her. Information must be treated on a 'need to know' basis, and not transmitted unnecessarily to others. In some cases, on defence projects, for example, consultants will be under an agreement to not divulge information to others. The responsibility placed upon the professional to act on behalf of others will be increased and his or her impartiality and independence become of great importance. He or she must be able to raise matters of concern with the client and if need be with higher authorities if there is serious concern about the consequences of some action. The professional does not, in his or her professional role, become a servant.

> One distinguishing feature of any profession is that membership in that profession entails an ethical obligation to temper one's selfish pursuit of economic success by adhering to standards of conduct that could not be enforced either by legal fiat or through the discipline of the market. Both the special privileges incident to membership in the profession, and the advantages those privileges give in the necessary task of earning a living, are the means to a goal that transcends the accumulation of wealth. That goal is public service.
>
> (US Supreme Court Judge Sandra Day Occoner)

The effect of 'the profession' on the role of the engineer

> The Engineer is a Mediator between the Philosopher and the Working Mechanic, and like an interpreter between two foreigners, must understand the language of both, hence the absolute necessity of possessing both practical and theoretical knowledge.
>
> (Henry Palmer. Inaugural address of the ICE, 2 January 1818)

If this definition of the engineer is accepted then the professional engineer, acting as such, can never truly be an employee. The true professional needs and retains the freedom to act in accordance with personal judgement, unbiased by the overriding needs of the employer.

In all of these roles, the engineer can act in different capacities: as a consultant, an employee, a manager and an employer. These roles can frequently produce conflicts of interest.

When the engineer becomes a manager, a further complication is added. The most important conflict is that involved in the decision-making processes. This arises from a conflict in loyalties. For example, it may be held that as a manager he or she has an overriding loyalty to the company, to its efficiency and profitability. As professionals, engineers have responsibilities towards the wider interests of safety, health and welfare of the public. When these criteria are applied to 'choices', conflict is inevitable, and the need for professional qualities becomes very important.

These qualities, together with the knowledge of the different levels of consequences of actions and of the nature of the society in which the actions are taking place, are essential if the engineer's decisions are to be clear and independent, and truly professional.

Engineering projects, particularly civil engineering projects, are multi-faceted endeavours. They are complex, unique, and built to strict deadlines of cost, time and quality; and are associated with high value and high uncertainty. They are invariably the result of team efforts, often involving 'inexperienced' clients. These constraints require detailed coordination and project management, a role often undertaken by the professional engineer.

> The function of the project manager is to foresee or predict as many of the dangers and problems as possible and to plan, organize and control activities so that the project is completed successfully in spite of the risks.
>
> (Locke 1992)

The engineer's technical skills must be augmented by other skills involving human relationships, administration, information management, programming, financial reporting and control and quality management. Managers are not usually considered as professionals in the full sense, being concerned with matters and standards that are more organization centred, whilst engineers are governed by standards beyond the organization, usually set by their training, codes of practice and institutional standards.

It may be assumed that the professional engineer serves society. This gives a particular social status, demanding autonomy in decision-making to ensure that judgement is not compromised by matters of expediency or finance. The engineer can be seen as a custodian of the concerns of many different groups of individuals or interests, including the environment and those who have no present voice, such as future generations.

The engineer has to survive, however, and the engineering company needs profit in order to survive. At the same time, the engineer cannot operate simply in a market-orientated way. The creative work of the engineer cannot be seen as simply a product. It involves a great deal of skill unavailable to the layperson. The professional engineer is not simply selling his or her services in the market, with the buyer being responsible for deciding between different products (caveat emptor – let the buyer beware), but is rather offering services, with the buyer needing to trust his or her skills and judgement (credat emptor – let the buyer trust).

The professional institution

> A Society for the general advancement of Mechanical Science and more particularly for promoting the acquisition of that species of knowledge which constitutes the profession of a Civil Engineer, being the art of directing the Great Sources of Power in Nature for the use and convenience of man.
>
> (Charter of the ICE 1828)

The virtue of integrity is not simply an individual one but applies to the professional institution and management groups as well: The profession as a whole has to be consistent and be able to relate values to practice. Without this the recognition of the profession as a body of experts concerned for public welfare will be eroded and with that will go the trust essential to the functioning of any professional relationship.

The existence of an institution of professionals is essential for the development and maintenance of professional virtues. It enables the individual professional to reflect upon his or her own integrity and to learn from the experience of others, transmitting the culture from generation to generation. It provides an external perspective that enables proper reflection and responsibility.

The institution can:

- *Enable* the professional development of moral awareness, skills, responsibility and identity, through codes, dialogues and training.
- *Ensure* its processes and organization are conducive to the development of moral responsibility.
- *Provide* support and the opportunity for professionals to work through decision-making and any conflicts of interest.
- *Regulate* the practice of the *individual professional.*

- *Play a major role in communicating* with the public.
- *Set standards* for admissions to institutions and for initial and continuing professional training.
- *Act* as a learned society, contributing to the advancement of science and technology of engineering.

The professional institution must avoid the moral ambiguity of acting as a body protecting the interests of its individual members – a trade union or a trade association – and must act as a body concerned about ensuring the highest standards of professional service to the community by regulating standards and criteria for membership. The institution itself has to develop a means of reflecting on its own ethos and ethical culture – this is a prime responsibility of the council of the institution, requiring regular reviews of the professional code and ensuring sound practice by all members.

The professional code

> All these trust in their hands; and everyone is wise in his work. Without these cannot a city be inhabited; and they shall not dwell where they will. . . . . But they will maintain the state of the world and all their desire is in the work of their craft.
>
> (Ecclesiasticus – Chap 38 v.31/32–34)

Professional codes enable ethical reflection and development, and are a means of developing the integrity of the professional body. The public declaration of the principles and views on responsibility and practice of a professional organization provide a benchmark against which the practice of the profession can be evaluated.

Ethical behaviour is an essential component in effective human societies. The aim is to enable collaboration between individual members, thereby improving the overall quality of life and fulfilment in the society, by creating an atmosphere of trust. The increasing awareness of social, international and global issues has expanded our awareness and is moving towards a sense of responsibility that extends far more widely than was the case when professional standards were first formulated.

The rules of the Institution of Civil Engineers focus mainly on the interests of clients and fellow professionals. It was felt necessary to extend the range of interests to consider the responsibilities of engineers towards the environment and to all who are affected by their work. The levels of responsibilities of all human beings, not only of engineers, can be formulated as ranging from individual 'goods', through family care and the interests of various social groups, such as professional organizations, to mankind as a whole and the general quality of our habitat. These were summarized as what is good for:

- The environment
- Mankind
- My associates (other professionals, citizens, countrymen)
- My family
- Myself

This is generally the basis of most institutional codes, as will be seen in the review later in this chapter.

What information should professional ethical codes contain?

The main issues that must be covered by professional codes are now fairly easy to identify. These must obviously include to who/whom and to what the professional owes responsibility, how members should behave and the consequences of breaking the code.

The code will probably need to recognize that members will join the particular profession in different roles. For example, if we take the case of the construction industry, members may be seen as employees, as employers and as designers with little or no part in the actual construction of the project, or as constructors with little or no part in the design. Each of these roles can have different duties and responsibilities.

The code should also ideally give guidance to those members who find themselves working in situations where they are subject to more than one allegiance – say an engineer working in the capacity of a company owner or manager.

Without attempting to put these issues into any order of importance it can be seen that the code will specifically need to address:

1. Responsibility to the profession.
2. Responsibility to oneself.
3. Responsibility to the employer, with the member acting as an employee.
4. Responsibility to the client.
5. Responsibility to the other individual members of the group or profession.
6. Responsibility to the community.

7. Responsibility to the environment.
8. Responsibility to other groups or professions.
9. In addition, the code will have to address issues concerning responsibilities of confidentiality – and probably cover whistleblowing.
10. Finally the code will also have to contain statements of how it will be determined if members have broken the institution's ethical guidelines – and the consequences if they are shown to have done so.

Investigating the contents of this list in more detail:

1. Responsibility to the profession
   This is essentially the accountability and responsibility of the member for his or her own professional performance, ensuring the integrity of the profession as a consequence. It involves maintaining and supporting the profession, and the exchange of ideas, expertise, learning and information with others in the group via the professional body; refusing to work in an unethical manner, consistent with the code; supporting the values of the professional body; bringing to their attention any ethical issues that need to be addressed as part of the code.
2. Responsibility to oneself
   This involves working within personal levels of competency consistent with training and qualifications and not claiming personal or professional capabilities beyond those actually attained; developing and maintaining personal development throughout working life; ensuring a proper balance between work and personal life.
3. Responsibility to the employer
   This involves being accountable and responsible for the performance of duties to an employer; acting with integrity and competence as long as they are consistent with the professional code; refusing to participate in unethical practices; recognizing the need and value of working cooperatively with other professionals or members of a multi-disciplinary team; respecting the unique contribution of each member and discipline.
4. Responsibility to the client
   This involves giving impartial and competent advice; reporting any conflicting areas of interest.
5. Responsibility to the other individual members of a team
   This involves issues of discrimination, diversity and harassment; recognition that competence for tasks may require advice or collaboration with experts in other professional fields, on a confidential basis.

6. Responsibility to the environment and the community
   This involves carrying out comprehensive and systematic reviews of the project with respect to both health and safety and the impact of the project on society and the environment – both immediate and in the long term (these reviews include justification, social and political factors and cost–benefit analyses); appraising and, where possible, involving the general public in the decision-making processes; making clear, objective and truthful public statements on these matters.
   Where confidentiality is required for defence, commercial or security reasons it is important to recognize responsibilities on behalf of the public and third-party interests.
7. Responsibility to other members, groups or professions
   This involves not attempting to supplant or maliciously injure the reputation or business of another person, engineer or professional.
8. Responsibilities of confidentiality – whistleblowing
   This involves respecting the confidentiality of clients and employers; following approved procedures for reporting and resolving whistleblowing issues.
9. Responsibility to follow and obey the code and the consequences if the rules are shown to have been broken
   This involves complying with institution rules and reporting infringements; supporting the institution in the expulsion of members found to have infringed ethical codes.

It can now be recognized that each code has the potential to become a comprehensive and lengthy document, with many sections and descriptions that themselves will be open to a variety of interpretations.

It is important, however, that the finished code does not become too complex or legalistic in style as it is intended to be read as everyday guidance to ethical problems.

The ways that various institutions have attempted to draft their codes to address these issues and ways that any individual or organization can prepare its own code will be reviewed later in the chapter.

Exercise

Examine the two cases noted below and identify the different responsibilities involved.

Case 4.1

Imagine that you are an experienced engineer. You occasionally pause to watch the progress of a construction site in the middle of a city where you work. A viewing place has been formed in the security fence specifically for members of the public. You notice a problem with a tower crane, which you feel could lead to a failure if not rectified. The failure is not likely to take place immediately but it appears to you that the crane could collapse and parts of the boom fall into the site and adjacent street unless some action is taken soon. Do you have any responsibility for action to prevent the possible accident? If so, to whom?

Case 4.2

You are used to working in the United Kingdom but this is the first time you have worked overseas in a developing country. You are disturbed to see the poor health and safety standards everywhere around you. Work would certainly not be allowed to be carried out under these conditions in the United Kingdom. What should you do?

The proliferation of engineering codes of ethics

Any review of existing engineering codes of ethics quickly uncovers a multitude of documents. The sheer number available suggests that every engineering institution and even branches of institutions have prepared a separate code for themselves.

This probably arises out of the diversity of the branches of engineering and the way engineers work. At one level an engineer may act as a consultant – an independent adviser investigating, designing and reporting on specific projects. Another engineer may be an employee of an organization not specifically concerned with engineering projects. They may be managers of engineering concerns or suppliers of goods and services. They may be designers, constructors or both. They may work in vastly different environments and cultures, in their home countries or largely overseas, operating under familiar laws and customs or in situations alien to them.

Overall, engineering can be seen as not one profession but as a family of professions loosely interrelated and developing over time as the scope and complexity of specialists, materials, techniques, environments and projects emerged. The Engineering Council UK (EC UK) attempts to regulate the engineering profession in the United Kingdom through the thirty-six engineering institutions that are licensed to place qualified members on their register. The highest grade of these professionals is that of Chartered Engineer, and their titles are protected by the Royal Charter. Registration is seen as desirable but not mandatory.

For many years, attempts have been made to prepare a single code for all engineers. The single code, it was hoped, would unify the many branches – but the radical independence of the members and the length and complexity of the code required to encompass the special circumstances, exceptions, diversity and many other factors have prevented this from happening. The result is that many codes exist, often differing quite radically in content and length. Many engineers find themselves bound by several different codes, sometimes with conflicting ideas, by virtue of working across boundaries of engineering disciplines. The EC UK has produced its own 'Guidelines for Institution Codes of Practice', which is included in Appendix A of this chapter. The general principles it contains are included in many of the UK institution codes.

It should be remembered that in many societies there is no legal restriction on anyone calling himself or herself an 'engineer' and practising accordingly. Whilst most professional institutions encourage registration and membership as part of an engineering career, many engineers do not become members – either because they do not meet the educational or training standard of the institutions or simply because they do not see the relevance of these bodies and do not wish to join them. It is reported that almost half of all American engineers do not belong to a professional association.

There are many difficulties in the current situation but these do not generally seem to cause major problems. The pragmatic engineer recognizes that these codes of ethics are moral rules. Often, the rules are not enforced by the organizations adopting them and those members violating them are unlikely to be censured and even less likely to be excluded from the organization. Often, the consequences of a moral or ethical lapse will be limited to the poor opinion of colleagues or to personal pangs of conscience. The engineer is more often likely to be concerned with justifying his or her conduct to those concerned rather than having to consider what some organization of engineers has to say about what he or she did or failed to do.

Existing codes – layout and content

Having studied the background to professional codes we are in a position to review the structure and contents of a selection of existing codes. There is a great proliferation of codes from almost every institution and professional body throughout the United Kingdom and the world and so in keeping with the spirit of this book we will confine ourselves to looking at engineering codes from the United Kingdom, the United States and Australia.

Appended to this chapter is a selection of seven such codes:

A. Guidelines for Institution Codes of Conduct. *Engineering Council* (EC UK))

B. Statement of Ethical Principles. *Royal Academy of Engineers* (RAE UK)

C. Code of professional conduct. *Institution of Civil Engineers* (ICE UK)

D. Code of Ethics. *American Society of Civil Engineers* (ASCE USA)

E. Code of Ethics. *The Institution of Engineers, Australia* (IEA Aus)

F. Code of Ethics. *Institute of Electrical and Electronics Engineers* (IEEE USA)

G. The California Contractor's Code of Ethics (CC USA)

Reference will also be made to other codes as appropriate.

An initial review shows that codes in A, B, C, D and E are broadly of the same form. F is a shorter and more businesslike code. G is included as a comparator – a 'nonprofessional' organization of contractors simply setting out their intentions in everyday dealings with their clients

EC UK suggests a code of practice that should be placed as a personal obligation to the members of the engineering institutions in the United Kingdom. Each member should act with integrity, in the public interest, and exercise all reasonable skill and care to fulfil the responsibilities included in the list.

RAE UK confines itself to essentially the same 'fundamental principles that guide an engineer in achieving the high ideals of professional life'. It provides a good example of an aspirational ethical code, working in broad principles and providing a good sense on conduct, virtues and attitudes.

ICE UK describes the rules of professional conduct and provides guidance notes for its members. This code presents a strong sense of community with legalistic undertones. It is based on an old document that has been rewritten and added to over the years. The format is one of a section on the rules of behaviour followed by a section on interpretation of the rules. The ICE code of professional conduct retains a strong feeling of an institution recognizing its responsibility and working as a force for the good of society.

ASCE USA describes the same intentions and responsibilities in a more modern fashion. Fundamental canons and fundamental principles are used and the document also provides guidelines to practice. It has a feeling of being carefully thought through. Its definitions are well thought out, and it distinguishes broad principles from guidelines.

IEA Aus contains sections on code of ethics and guidance for members. This document reads like a modern serious attempt to demystify the field of professional ethics and guide its members in the current environment. There are similar references to the ethical rules and guidelines but contemporary issues such as expert witness behaviour and whistleblowing are explicitly covered.

IEEE USA is of a slightly different format. It contains a straightforward charter of ten points that commit its members to 'the highest ethical and professional conduct'.

CC USA is included in this review as a comparator. This is a code of ethics prepared by a group of contractors solely for the purpose of setting out rules of behaviour in dealing with their clients. There is no inherent view of their responsibilities to society. It can be regarded as a good straightforward 'nonprofessional' code.

A detailed examination of the professional codes reveals a desire to promote the common good on the basis of ethical behaviour, competent performance, innovative practice, engineering excellence, equality, social justice and sustainable development.

All institutions covered by this review aspire to follow these principles. Importantly, each code carefully spells out the particular institution's responsibility to society and it is this aspect we will now investigate in more detail.

The professional code and responsibility to society

One of the key elements of a professional engineering code has been shown to be the attempt to serve the higher interests of society rather than solely the requirements of the client. It is interesting to compare how the five various governing bodies of the institutions in the United Kingdom attempt to set out this goal.

Examples of specific requirements relevant to public duties in each are:

- A member shall at all times so order his conduct as to safeguard the public interest, particularly in matters of Health & Safety and the Environment. – *Institution of Mechanical Engineers, UK Rule 33.6.*
- A member shall at all times take all reasonable care to ensure that their work and the products of their work constitute no avoidable danger of death or injury or ill-health to any person; and take all reasonable steps to avoid waste of natural resources, damage of the environment, and wasteful damage or destruction of the products of human skill and industry. – *Institution of Electrical Engineers, UK Rules 1 and 2.*
- Every corporate member shall safeguard the public interest in matters of Health & Safety and otherwise. – *Institution of Structural Engineers, UK Code of Conduct, Rule I.*
- Members of the institution in their responsibility to the profession shall have full regard to the public interest. – *Institution of Structural Engineers, UK Rules of Conduct, Rule I.*

- A member shall have full regard for the public interest, particularly in relation to the environment and to matters of Health & Safety. – *Institution of Civil Engineers, UK Rules of Professional Conduct, Rule 3.*

The same concern for social and public environment is there in American codes:

> Engineers uphold and advance the integrity, honor and dignity of the engineering profession by:
>
> 1. Using their knowledge and skill for the enhancement of human welfare and the environment.
> 2. Being honest and impartial and serving with fidelity the public, their employers and clients.
> 3. Striving to increase the competence and prestige of the engineering profession.
> 4. Supporting the professional and technical societies of their disciplines. *American Society of Civil Engineers USA Fundamental Principles Code of Ethics.*
>
> Engineers shall hold paramount the safety, health and welfare of the public and shall strive to comply with the principles of sustainable development in the performance of their professional duties. *Fundamental Canon. American Society of Civil Engineers USA Code of Ethics.*

All these codes recognize the professional retaining the freedom to act in accordance with personal judgement, recognizing his or her responsibilities to the employer but acting in a manner unbiased by the overriding needs of the employer and always having responsibility to the public as the paramount duty.

Engineering codes and the law

Our review has highlighted the focus on the role of the engineer as a person serving wider society as well as any clients. This is the primary factor that differentiates engineering codes from almost all other professions.

It is interesting to pursue this idea further and investigate whether this difference – so carefully spelt out in so many codes – has any true meaning under the law.

John Uff (2005) argues that engineers have no special status under the law and simply make contractual arrangements for whatever work they may undertake. Professional codes of conduct are not usually incorporated into these contracts of engagements. The inference is that the general duty to uphold the public interest, which is set out in various clauses in almost all of the institutional rules and codes, is not enforceable directly by law. The onus is accordingly placed on the institutions, which publish the codes and rules, to enforce compliance. This is notionally done by depriving those who break the rules of their professional membership.

These procedures are more easily observed in other nonengineering professions. Most particularly, in medicine, we often read of a medical practitioner being struck off the General Medical Council register and being prevented from practising as a result of malpractice. A similar public history of engineering disciplinary procedures is not there, probably because, in the main, there are no similar requirements for engineers to be registered before they can practise and no enforceable restrictions on practise without registration.

Guidance for writing ethical codes

There are three basic elements in writing an ethical code (often referred to as code of practice):

- Determine the purpose of the code. Is it to modify, regulate or control behaviour, articulate and maintain standards or to create a new and better ethos? There may be elements of all of these in a code.
- Articulate any mission or vision statements, ensuring that the code embodies the underlying principles. Hence, aspirations can be differentiated from standards. If there are standards then these must be enforced. What are the means of compliance?
- Group involvement. Carefully reflect upon for whom the code is to be written. Would it be best to involve the members who will eventually be governed by the code to assist or even write the code themselves? What are the beliefs and values of the members? These must be at the heart of the code if it is to succeed. Think of how the code will be implemented, published, updated and how it will be revised when, in the light of future use, it needs to be amended.

However good an ethical code is, it cannot be the only means of ethical response, and therefore must not be seen as a set of prescriptions. It provides guidelines, a means of reminding the professional about core shared values, responsibilities, standards and practices of the profession. At its root it should enable the professional engineer to develop his or her autonomy and responsibility, in order to respond appropriately to the context. Hence, the engineer also has to be familiar with basic principles, the ethical method, and to embody, along with the wider profession, the core moral virtues noted in Chapter 3.

Exercise

Look at the code of your professional engineering institution. Consider how far this ties in with the basic values of your profession. Does it tie in with mission and value statements? Does it give you effective guidance? How would you develop it to make it more effective?

Is it possible to have just one ethical code for all engineers? Consider what the main points of such a code would be.

Conclusions

This chapter has investigated professional ethical codes with a view to reviewing the purpose and content of codes and how they are used by the members and institutions for whom they were primarily prepared.

It has been shown that many organizations and institutions have written their own codes of ethics and that some of them have been in use for very long periods of time.

Engineering professions have been amongst this group and have always sought to clearly identify what sets them apart as a 'profession' from other groups and even other professions. Engineers have identified themselves as a group whose works are able to influence the quality of life of large sections of the population and even whole societies, who, for the main part, may not understand what the engineer does. The engineers therefore see themselves as working for clients and, while fulfilling their engineering contractual obligations, having a higher level of responsibility that, as Occoner noted of all professions, is 'public service'.

Most codes prepared by professional engineering institutions and organizations reflect this primary obligation of the safety, health and welfare of the public.

When this responsibility is viewed from a legal standpoint, however, it is evident that it has little or no meaning under the law. Engineers have no special status under the law and make normal contractual arrangements for work they undertake. Professional codes of conduct are not incorporated into these contracts of engagements. The general duty to uphold the public interest is therefore a voluntary responsibility that the engineer, and the professions of engineering, has chosen to adopt. The institutions and organizations that have incorporated these responsibilities into their codes of ethics are therefore the bodies responsible for their enforcement.

However, under UK law and under the laws of many other countries, any individual may call himself or herself an engineer and practise as such without having the training, education or the requirement to join one of the professional institutions. The result is that these bodies find it almost impossible to enforce their carefully formulated ethical codes. Until they acquire the status and power of an institution such as the General Medical Council, which licences and controls the education, training, practice and behaviour of its members, there will be no major changes in this situation.

The loose arrangement of control provided by the engineering institutions has encouraged the proliferation of ethical codes operated by the professions. Without one powerful body with the capacity to licence engineers, each separate institution will continue to formulate and maintain its own rules and regulations, ultimately to the detriment of the profession as a whole. The question we pose to the institutions is, how can we maintain their autonomy and, at the same time, have an ethical framework for all engineers that can have more assurance of compliance? Another way forward is to have an agreed core general code that remains the same for all the institutions, with the addition of regulations about issues that are particular to institutions. Perhaps Appendix A or B might fulfil such a core code.

This issue underlines the point that ethical meaning in professions and how it is handled by the institutions is, and should be, a matter of ongoing debate.

References

May, W. (1985). Adversarialism in America and in the professions. In *The End of Professionalism?* [Occasional Paper No. 6.] pp. 5–19. Edinburgh: Centre for Theology and Public Issues, University of Edinburgh.

Uff, J. (2005). *Engineering Ethics: Do Engineers Owe Duties to the Public*? London: Royal Academy of Engineers.

Appendix A

Engineering Council UK (EC UK)

Guidelines for institution codes of conduct

The code of professional conduct of each nominated engineering institution should place a personal obligation on its members to act with integrity, in the public interest, and to exercise all reasonable professional skill and care to:

1. Prevent avoidable danger to health or safety.

2. Prevent avoidable adverse impact on the environment.

3. A Maintain their competence.
   B Undertake only professional tasks for which they are competent.
   C Disclose relevant limitations of competence.

4. **A** Accept appropriate responsibility for work carried out under their supervision.
   B Treat all persons fairly, without bias, and with respect.
   C Encourage others to advance their learning and competence.

5. **A** Avoid where possible real or perceived conflict of interest.
   B Advise affected parties when such conflicts arise.

6. Observe the proper duties of confidentiality owed to appropriate parties.

7. Reject bribery.

8. Assess relevant risks and liability, and if appropriate hold professional indemnity insurance.

9. Notify the institution if convicted of a criminal offence or upon becoming bankrupt or disqualified as a company director.

10. Notify the institution of any significant violation of the institution's code of conduct by another member.

Appendix B

Royal Academy of Engineering (RAE UK)

Statement of ethical principles

The decisions and actions of engineers have a profound impact on the world we live in, and society at large. Making a clear and public commitment to operating with integrity and honesty is essential to create a greater level of trust and confidence, and a positive perception of the engineering profession.

The Royal Academy of Engineering, in collaboration with Engineering Council (UK) and a number of the leading professional engineering institutions, has created a Statement of Ethical Principles to which it believes all professional engineers and related bodies should subscribe.

Accuracy and veracity
Honesty and integrity
Engineering is the knowledge required, and the process applied, to conceive, design, make, build, operate, sustain, recycle or retire, something of significant technical content for a specific purpose – a concept, a model, a product, a device, a process, a system, a technology.

Engineers are individuals who apply this creative process. Professional engineers work to enhance the welfare, health and safety of all whilst paying due regard to the environment and the sustainability of resources. They have made personal and professional commitments to enhance the well-being of society through the exploitation of knowledge to create new things and the management of creative teams.

This statement of ethical principles establishes the standard which the members of the engineering profession adopt to regulate their working habits and relationships. The values on which it is based should apply whether or not an engineer is acting in a professional capacity.

There are four fundamental principles which guide an engineer in achieving the high ideals of professional life. These express the beliefs and values of the profession and are amplified below. To aid interpretation in practical situations guidance notes will be provided.

Statement of ethical principles

Professional engineers have a duty to ensure that they acquire and use wisely and faithfully all knowledge relevant to the engineering skills needed in their work in the service of others; they should:

- Act with care and competence in all matters relating to duties.
- Maintain up-to-date knowledge and skills and assist their development in others.
- Perform services only in areas of current competence.
- Not knowingly mislead, or allow others to be misled, about engineering matters.
- Present and review engineering evidence, theory and interpretation honestly, accurately and without bias and quantify all risks.
- There are fundamental common values that bind all humanity together: the profession derives its ultimate value from people. Accordingly, all dealings with others should be conducted with fairness and honesty and professional engineers should accord the highest importance to freedom of choice, equality of opportunity and social justice; they should:
    - Be alert to the ways in which their duties derive from and affect the work of other people; respect the rights and reputations of others.
    - Avoid deceptive acts and take steps to prevent corrupt practices and professional misconduct; declare conflicts of interest.
    - Reject bribery or improper influence.
    - Act for each employer or client in a reliable and trustworthy manner.

Respect for life, law and the public good
Responsible leadership: listening and informing
In making choices professional engineers should give due weight to all relevant law, facts and guiding principles and to the public interest; they should:

- Ensure that all work is lawful and justified.
- Minimise and justify any adverse effect on wealth creation, the natural environment and social justice whilst ensuring that all developments meet the needs of the present without compromising the ability of future generations to meet their own needs.
- Act honourably, responsibly and lawfully so as to uphold the reputation, standing and dignity of the profession.

Professional engineers should exercise high standards of leadership in the exploitation and management of technology. They hold a privileged and trusted position in society and have a duty to ensure that their position is not used to the benefit of personal or sectional interests or to the detriment of the wider community but is seen to reflect public concern; they should:

- Identify and be aware of the issues that engineering raises for society; listen to the aspirations and concerns of others.
- Lead in promoting public awareness and understanding of the impact and benefits of engineering achievements.
- Issue public statements only in an objective and truthful manner.

The Royal Academy of Engineering
The Royal Academy of Engineering
29 Great Peter Street, London SW1P 3LW
Tel: 020 7227 0500 Fax: 020 7233 0054 www.raeng.org.uk
The Royal Academy of Engineering promotes excellence in the science, art and practice of engineering.
Registered charity number 293074

Appendix C

Institution of Civil Engineers (ICE UK)

Code of professional conduct

Contents

Introduction

The purpose of the code of professional conduct

The ICE has provided a code of professional conduct to lay down, both for its members and for the general public, the ethical standards by which its members should abide. The code will apply to all its members, irrespective of their grade, the professional role they fulfil, and the countries in which they practise. The code contains, first of all, the specific rules of professional conduct to which members must adhere. The rules cover, in plain language, those basic things that members must do. Where appropriate, there are guidance notes to help members interpret and apply the rules (there are no guidance notes for Rules 4 and 6). The guidance notes are not exhaustive and do not cover all contingences. However, the guidance notes do cover the main areas where instances of improper conduct have arisen. An act which seriously breaches one of the guidance notes will probably breach the rule of professional conduct to which that guidance note applies.

In the course of their careers members may undertake roles that remove them from direct involvement with engineering matters. This does not, however, remove from them the obligation, as members of the ICE, to act in accordance with the rules of professional conduct in all aspects of their professional lives.

Advice to members

The ICE has published a separate document entitled 'Advice on Ethical Conduct'. This advice is to help members by providing more information on what the institution considers to be ethical conduct. A member will not be considered to have breached the rules of professional conduct solely because he or she has not followed the 'Advice on Ethical Conduct'.

The ICE is anxious that any members who are troubled by an ethical problem, particularly if they are worried about breaching the rules of professional conduct, should be able to discuss their concerns with one or more of the senior members of the ICE. Any members who wish to do this should contact the Professional Conduct Department so that arrangements can be made.

The duty to behave ethically

The duty upon members of the ICE to behave ethically is, in effect, the duty to behave honourably; in modern words, 'to do the right thing'. At its most basic, it means that members should be truthful and honest in dealings with clients, colleagues, other professionals, and anyone else they come into contact with in the course of their duties. Being a member of the ICE is a badge of probity and good faith, and members should do nothing that in any way could diminish the high standing of the profession. This includes any aspect of a member's personal conduct which could have a negative impact upon the profession.

Members of the ICE should always be aware of their overriding responsibility to the public good. A member's obligations to the client can never override this, and members of the ICE should not enter undertakings which compromise this responsibility. The 'public good' encompasses care and respect for the environment, and for humanity's cultural, historical and archaeological heritage, as well as the primary responsibility members have to protect the health and well-being of present and future generations.

Institution of Civil Engineers -D 2 - **Date** 9 July 2004 **Revision** 0

The rules of professional conduct

1. All members shall discharge their professional duties with integrity.
2. All members shall only undertake work that they are competent to do.
3. All members shall have full regard for the public interest, particularly in relation to matters of health and safety, and in relation to the well-being of future generations.

4. All members shall show due regard for the environment and for the sustainable management of natural resources.
5. All members shall develop their professional knowledge, skills and competence on a continuing basis and shall give all reasonable assistance to further the education, training and continuing professional development of others.
6. All members shall:

   a. notify the institution if convicted of a criminal offence;
   b. notify the institution upon becoming bankrupt or disqualified as a company director;
   c. notify the institution of any significant breach of the rules of professional conduct by another member.

Institution of Civil Engineers -D 3 - **Date** 9 July 2004 **Revision** 0

Guidance notes on the interpretation and application of the rules of professional conduct

Rule 1: All members shall discharge their professional duties with integrity

The manner in which members could breach this rule might include the following:

- Failing to carry out their professional duties with complete objectivity and impartiality.
- Failing to declare conflicts of interest.
- Failing to have due regard to their duty of confidence in relation to all parties with whom they have dealings as part of their professional duties.
- Failing to have due regard to their duty of care to clients, especially lay clients in domestic or small works engagements.
- Failing to put terms of the engagement in writing and failing to state the fees to be charged; whenever practicable, these should be issued to the client before the project is begun.

- Members who do not carry appropriate insurance, either personally or through their employers, failing to advise their clients of the position before accepting the engagement. Members should take all reasonable steps to ensure that their prospective clients understand the extent to which they are covered by appropriate insurance.
- Failing to show due consideration for other colleagues and for other persons with whom they have dealings in the course of their professional duties. Members must treat all persons without bias and with respect.
    - Members must not, directly or indirectly, improperly attempt to supplant another person, and should take all reasonable steps to establish that any previous engagement in relation to the project they are to enter into has been terminated.
    - When commenting on another person's work, members must advise that person of their involvement, except for routine or statutory checks or where the member's client or employer requires confidentiality.
    - When competing with other professionals, and when taking actions likely to adversely affect the professional or business interests of another person, members must not act maliciously or recklessly.
- Members who are responsible for the work of others or who have a management responsibility for other persons failing to take responsibility for the work produced. Members should ensure that they have the knowledge and expertise to effectively oversee the work for which they are accountable.
- Having any form of involvement, whether direct or indirect, and whether for the benefit of the member, the member's employer, or a third party, in bribery, fraud, deception and corruption. Members should be especially rigorous when operating in countries where the offering and accepting of inducements and favours, or the inflation and falsification of claims, is endemic.
- When acting as expert witnesses, failing to ensure that the testimony they give is both independent and impartial. In such a role, members must be mindful that their prime duty is to the court or tribunal, not to the client who engaged them to give evidence, and they should not give any professional opinion that does not accurately reflect their honest professional judgement or belief. To do otherwise would not only place members in danger of perjury but would clearly breach the requirement in the rules of professional conduct to discharge their professional duties with integrity.

Institution of Civil Engineers -D 4 - **Date** 9 July 2004 **Revision** 0

Rule 2: All members shall only undertake work that they are competent to do.

Members should be competent in relation to every project that they undertake. They should ensure that, having regard to the nature and extent of their involvement in a project, they have the relevant knowledge and expertise. Where appropriate, this may include access to the knowledge and experience of others, or access to other relevant sources of knowledge, in addition to the member's own knowledge and experience. In so doing, they should pay due regard to the laws on copyright and other rights of intellectual property. They must disclose, where appropriate, any relevant limitations upon their competence.

Rule 3: All members shall have full regard for the public interest, particularly in relation to matters of health and safety, and in relation to the well-being of future generations.

The manner in which members could breach this rule might include the following:

- Failing to take all reasonable steps to protect the health and safety of members of the public and of those engaged in the project, during the construction and the operation and maintenance stages. 'Reasonable steps' will include obeying all legislation relating to health and safety but may extend beyond that to all situations in which there is inadequate statutory provision. Members must not enter into any contracts which compromise this overall responsibility.
- Where new or under-tested materials or methods are being used, failing to pay a reasonable level of attention to the public safety implications, and failing to have regard to the possibility that performance of the materials or methods may be worse than forecast. The use of such materials or methods and the risks involved must be drawn specifically to the client's notice.

Other matters relating to this rule would include:

- Producing competitive bids should not result in the inappropriate exposure to hazard of any person at any time. Members have a duty, as far as is reasonable, to keep abreast of emerging hazards, and to inform interested parties accordingly.
- Members must take all reasonable steps to avoid preventable disasters and should act in accordance with relevant good practice; for example, for UK-based members this will be the Royal Academy of Engineering Guidelines for Warnings of Preventable Disasters. If members are in any doubt about the action they should take, they should seek the advice of the institution.

- Members should take account of the broader public interest – the interests of all stakeholders in any project must be taken properly into account, including the impact on future generations. This must include regard for the impact upon the society and quality of life of affected individuals, groups or communities, and upon their cultural, archaeological and ethnic heritage, and the broader interests of humanity as a whole.

Rule 5: All members shall develop their professional knowledge and skills on a continuing basis and shall give all reasonable assistance to further the education, training and continuing professional development of others.

- All members have a duty to improve and update technical knowledge, and to keep abreast of relevant developments, including new or changed statutory provisions.

Institution of Civil Engineers -D 5 - **Date** 9 July 2004 **Revision** 0

- Every member has a duty to be proactive in the training and continuing professional development of others, especially those for whom the member has line management responsibility.

Institution of Civil Engineers -D 6 - **Date** 9 July 2004 **Revision** 0
onlineethics.org *The Online Ethics Center for Engineering and Science*

Appendix D

American Society of Civil Engineers (ASCE USA)

Code of ethics

Fundamental principles

Engineers uphold and advance the integrity, honor and dignity of the engineering profession by:

1. Using their knowledge and skill for the enhancement of human welfare and the environment.
2. Being honest and impartial and serving with fidelity the public, their employers and clients.
3. Striving to increase the competence and prestige of the engineering profession.
4. Supporting the professional and technical societies of their disciplines.

Fundamental cannons

1. Engineers shall hold paramount the safety, health and welfare of the public and shall strive to comply with the principles of sustainable development in the performance of their professional duties.
2. Engineers shall perform services only in areas of their competence.
3. Engineers shall issue public statements only in an objective and truthful manner.
4. Engineers shall act in professional matters for each employer or client as faithful agents or trustees, and shall avoid conflicts of interest.
5. Engineers shall build their professional reputation on the merit of their services and shall not compete unfairly with others.
6. Engineers shall act in such a manner as to uphold and enhance the honour, integrity, and dignity of the engineering profession.

7. Engineers shall continue their professional development throughout their careers, and shall provide opportunities for the professional development of those engineers under their supervision.

Guidelines to practice under the fundamental cannons of ethics

Canon 1

Engineers shall hold paramount the safety, health and welfare of the public and shall strive to comply with the principles of sustainable development in the performance of their professional duties.

1. Engineers shall recognize that the lives, safety, health and welfare of the general public are dependent upon engineering judgments, decisions and practices incorporated into structures, machines, products, processes and devices.
2. Engineers shall approve or seal only those design documents reviewed or prepared by them, which are determined to be safe for public health and welfare in conformity with accepted engineering standards. Engineers whose professional judgment is overruled under circumstances where the safety, health and welfare of the public are endangered, or the principles of sustainable development ignored, shall inform their clients or employers of the possible consequences.
3. Engineers who have knowledge or reason to believe that another person or firm may be in violation of any of the provisions of canon 1 shall present such information to the proper authority in writing and shall cooperate with the proper authority in furnishing such further information or assistance as may be required.
4. Engineers should seek opportunities to be of constructive service in civic affairs and work for the advancement of the safety, health and well-being of their communities, and the protection of the environment through the practice of sustainable development.
5. Engineers should be committed to improving the environment by adherence to the principles of sustainable development so as to enhance the quality of life of the general public.

Canon 2

Engineers shall perform services only in areas of their competence.

1. Engineers shall undertake to perform engineering assignments only when qualified by education or experience in the technical field of engineering involved.
2. Engineers may accept an assignment requiring education or experience outside of their own fields of competence, provided their services are restricted to those phases of the project in which they are qualified. All other phases

of such a project shall be performed by qualified associates, consultants or employees.

3. Engineers shall not affix their signatures or seals to any engineering plan or document dealing with subject matter in which they lack competence by virtue of education or experience or to any such plan or document not reviewed or prepared under their supervisory control.

Canon 3

Engineers shall issue public statements only in an objective and truthful manner.

1. Engineers should endeavor to extend the public knowledge of engineering and sustainable development, and shall not participate in the dissemination of untrue, unfair or exaggerated statements regarding engineering.
2. Engineers shall be objective and truthful in professional reports, statements, or testimony. They shall include all relevant and pertinent information in such reports, statements, or testimony.
3. Engineers, when serving as expert witnesses, shall express an engineering opinion only when it is founded upon adequate knowledge of the facts, upon a background of technical competence, and upon honest conviction.
4. Engineers shall issue no statements, criticisms, or arguments on engineering matters which are inspired or paid for by interested parties, unless they indicate on whose behalf the statements are made.
5. Engineers shall be dignified and modest in explaining their work and merit, and will avoid any act tending to promote their own interests at the expense of the integrity, honor and dignity of the profession.

Canon 4

Engineers shall act in professional matters for each employer or client as faithful agents or trustees, and shall avoid conflicts of interest

1. Engineers shall avoid all known or potential conflicts of interest with their employers or clients and shall promptly inform their employers or clients of any business association, interests, or circumstances which could influence their judgment or the quality of their services.
2. Engineers shall not accept compensation from more than one party for services on the same project, or for services pertaining to the same project,

unless the circumstances are fully disclosed to and agreed to by all interested parties.

3. Engineers shall not solicit or accept gratuities, directly or indirectly, from contractors, their agents, or other parties dealing with their clients or employers in connection with work for which they are responsible.
4. Engineers in public service as members, advisors, or employees of a governmental body or department shall not participate in considerations or actions with respect to services solicited or provided by them or their organization in private or public engineering practice.
5. Engineers shall advise their employers or clients when, as a result of their studies, they believe a project will not be successful.
6. Engineers shall not use confidential information coming to them in the course of their assignments as a means of making personal profit if such action is adverse to the interests of their clients, employers or the public.
7. Engineers shall not accept professional employment outside of their regular work or interest without the knowledge of their employers.

Canon 5
Engineers shall build their professional reputation on the merit of their services and shall not compete unfairly with others.

1. Engineers shall not give, solicit or receive, either directly or indirectly, any political contribution, gratuity, or unlawful consideration in order to secure work, exclusive of securing salaried positions through employment agencies.
2. Engineers should negotiate contracts for professional services fairly and on the basis of demonstrated competence and qualifications for the type of professional service required.
3. Engineers may request, propose or accept professional commissions on a contingent basis only under circumstances in which their professional judgments would not be compromised.
4. Engineers shall not falsify or permit misrepresentation of their academic or professional qualifications or experience.
5. Engineers shall give proper credit for engineering work to those to whom credit is due, and shall recognize the proprietary interests of others. Whenever possible, they shall name the person or persons who may be responsible for designs, inventions, writings or other accomplishments.

6. Engineers may advertise professional services in a way that does not contain misleading language or is in any other manner derogatory to the dignity of the profession. Examples of permissible advertising are as follows:

   - Professional cards in recognized, dignified publications, and listings in rosters or directories published by responsible organizations, provided that the cards or listings are consistent in size and content and are in a section of the publication regularly devoted to such professional cards.
   - Brochures which factually describe experience, facilities, personnel and capacity to render service, providing they are not misleading with respect to the engineer's participation in projects described.
   - Display advertising in recognized dignified business and professional publications, providing it is factual and is not misleading with respect to the engineer's extent of participation in projects described.
   - A statement of the engineers' names or the name of the firm and statement of the type of service posted on projects for which they render services.
   - Preparation or authorization of descriptive articles for the lay or technical press, which are factual and dignified. Such articles shall not imply anything more than direct participation in the project described.
   - Permission by engineers for their names to be used in commercial advertisements, such as may be published by contractors, material suppliers, etc., only by means of a modest, dignified notation acknowledging the engineers' participation in the project described. Such permission shall not include public endorsement of proprietary products.

7. Engineers shall not maliciously or falsely, directly or indirectly, injure the professional reputation, prospects, practice or employment of another engineer or indiscriminately criticize another's work.
8. Engineers shall not use equipment, supplies, laboratory or office facilities of their employers to carry on outside private practice without the consent of their employers.

Canon 6
Engineers shall act in such a manner as to uphold and enhance the honour, integrity, and dignity of the engineering profession.

1. Engineers shall not knowingly act in a manner which will be derogatory to the honor, integrity, or dignity of the engineering profession or knowingly engage in business or professional practices of a fraudulent, dishonest or unethical nature.

Canon 7

Engineers shall continue their professional development throughout their careers, and shall provide opportunities for the professional development of those engineers under their supervision.

1. Engineers should keep current in their specialty fields by engaging in professional practice, participating in continuing education courses, reading in the technical literature and attending professional meetings and seminars.
2. Engineers should encourage their engineering employees to become registered at the earliest possible date.
3. Engineers should encourage engineering employees to attend and present papers at professional and technical society meetings.
4. Engineers shall uphold the principle of mutually satisfying relationships between employers and employees with respect to terms of employment including professional grade descriptions, salary ranges, and fringe benefits.

1. As adopted September 2, 1914, and most recently amended November 10, 1996.
2. The American Society of Civil Engineers adopted The Fundamental Principles of the ABET Code of Ethics of Engineers as accepted by the Accreditation Board for Engineering and Technology, Inc. (ABET). (By ASCE Board of Direction action April 12–14, 1975)
3. In November 1996, the ASCE Board of Direction adopted the following definition of sustainable development: 'Sustainable development is the challenge of meeting human needs for natural resources, industrial products, energy, food, transportation, shelter, and effective waste management while conserving and protecting environmental quality and the natural resource base essential for future development.'

Appendix E

The Institution of Engineers, Australia (IEA Aus)

Code of ethics

Preamble

Engineering is a creative process of synthesising and implementing the knowledge and experience of humanity to enhance the welfare, health and safety of all members of the community, with due regard to the environment in which they live and the sustainability of the resources employed. It involves a diversity of related functions ranging from the development and application of engineering science through to the management of engineering works. The members of the Institution of Engineers, Australia, are bound by a common commitment to promote engineering and facilitate its practice for the common good based upon shared values of:

- Ethical behaviour
- Competent performance
- Innovative practice
- Engineering excellence
- Equality of opportunity
- Social justice
- Unity of purpose
- Sustainable development

The community places its trust in the judgement and integrity of members to pursue the above values and to conduct their activities in a manner that places the best interests of the community above those of personal or sectional interests. The code of ethics provides a statement of principles which has been adopted by the council of the institution as the basis upon which members shall conduct their activities in order to merit community trust. It is also the framework from which rules of conduct may be developed.

The tenets of the code of ethics embrace principles which are immutable; however, changing community perceptions require that periodic reviews of the tenets be conducted. The 1994 issue of the code represents a significant revision of the text to reflect the changes in expectations of the community and the broader role of the institution in community affairs. The code is accompanied by a section which provides more specific guidance on the application of the principles to meet community expectations. Members are required to abide by the tenets as part of their commitment to participate in the affairs of the institution. Accordingly, all members are required to give active support to the proper regulation of qualifications, employment and practice in engineering.

Members acting in accordance with this code will have the support of the institution. The manner and extent of the support will be determined by the council of the institution on the merits of each case.

The code of ethics

The members of the Institution of Engineers, Australia, are committed to the cardinal principles of the code:

- To respect the inherent dignity of the individual.
- To act on the basis of a well-informed conscience.
- To act in the interest of the community and to uphold its tenets.

The tenets of the code of ethics are:

1. Members shall at all times place their responsibility for the welfare, health and safety of the community before their responsibility to sectional or private interests, or to other members.
2. Members shall act in order to merit the trust of the community and membership in the honour, integrity and dignity of the members and the profession.
3. Members shall offer services or advise on or undertake engineering assignments only in areas of their competence and shall practise in a careful and diligent manner.
4. Members shall act with fairness, honesty and in good faith towards all in the community, including clients, employers and colleagues.
5. Members shall apply their skill and knowledge in the interest of their employer or client for whom they shall act as faithful agents or advisers, without compromising the welfare, health and safety of the community.

6. Members shall take all reasonable steps to inform themselves, their clients and employers and the community of the social and environmental consequences of the actions and projects in which they are involved.
7. Members shall express opinions, make statements or give evidence with fairness and honesty and on the basis of adequate knowledge.
8. Members shall continue to develop relevant knowledge, skill and expertise throughout their careers and shall actively assist and encourage those under their direction to do likewise.
9. Members shall not assist, induce or be involved in a breach of these tenets and shall support those who seek to uphold them.

Guidance for members

The code of ethics establishes the standard which the members of the institution adopt to regulate their working habits and relationships. The principles on which it is based should apply equally in the members' personal lives.

The code is structured on two tiers covering cardinal principles which guide all behaviour and the linked tenets based on more specific principles to which the members of the institution ascribe. The following section amplifies the essence of the cardinal principles and identifies the specific principles which underlie the tenets. Subsequent sections provide interpretations of the tenets as they apply to practices and situations in which members may find a need for ethical guidance.

Principles

The cardinal principles express the beliefs and values of the members of the institution based on the recognition that:

a. There are fundamental common ties that bind all humanity together and that our institutions derive their ultimate value from people. Accordingly, our expectations and performance in dealing with others should be conducted with fairness and honesty and members should accord the highest importance to freedom of choice, equality of opportunity and social justice.
b. In the face of conflicting requirements, the content and quality of our choices are finally a matter of personal responsibility, and that in coming to any decision members should give due weight to all relevant facts and guiding principles as far as they can be ascertained.
c. Members hold a privileged and trusted position in the community. Members have a duty to ensure that this position is not used for personal or sectional interests to the detriment of the wider community.

The tenets express the shared commitment of the members to act in a manner which upholds the cardinal principles and are based on the more specific principles expressed by:

- Behaviour engendering community trust.
- Risk being managed in the interest of the community.
- The community having the right to be informed.
- A responsibility of service to clients or employers.
- Practice being in accord with sustainability and environmental principles.
- Fairness in dealing with others.
- Relationships being on an open and informed basis.
- Opinions or evidence being a balanced and full representation of the truth.
- Knowledge being current to serve best the interests of the community, employers and clients.
- Awareness of the consequences of one's own actions.
- A shared responsibility to uphold the tenets.

Interpretation

The nine tenets of the code are of necessity couched in broad terms. The comments which follow are provided to expand on and discuss some of the more difficult and interrelated components of the code, without narrowing its focus. They are provided to assist members to understand the code. However, they are not part of the code. The specific interpretations and guidance on ethical obligations raised should not be seen as limiting the scope of the code, nor should they be seen as exhaustive. A breach of the code of ethics occurs when a member acts contrary to one of the nine tenets judged on the circumstances of the case and not on the emphasis of the interpretations.

A member or other person requiring further guidance should obtain it from a divisional office or the national office of the institution.

The community

The commitment of members to act in the interest of the community is fundamental to the ethical values of the profession. The term community should be interpreted in its widest context to comprise all groups in society, including the member's own workplace. Members' obligation to the welfare, health and safety of the community involves the application of sound engineering judgement based on experience and relevant analysis to arrive at the appropriate balance of considerations which must apply in any given situation. Protection of the environment is both a short-term and a long-term concern of the community and needs to be considered by members at all times. Members' obligations extend to taking reasonable steps to understand the consequences of their own actions and the actions of those with or for whom they are working.

Members:

a. Shall work in conformity with accepted engineering and environmental standards and a manner which does not jeopardise the public welfare, health or safety.

b. Shall endeavour at all times to maintain engineering services essential to public health and safety.

c. Shall have due regard to requirements for the health and safety of the workforce.

d. Shall give due weight to the need to achieve sustainable development and to conserve and restore the productive capacity of the earth.

e. Shall endeavour to ensure that information provided to the public is relevant and in a readily understood form.

f. Shall avoid assignments taken on behalf of clients or employers that are likely to create a conflict of interest between the member or their clients or employers and the community.

g. Shall not use association with other persons, corporations, or partnerships to conceal unethical acts.

h. Shall not involve themselves with any practice that they know to be of a fraudulent, dishonest or criminal nature, whether involving engineering activities or otherwise. Successful prosecution before a court for any such action may be deemed to be a breach of the code of ethics.

Areas of competence and description of qualifications

Members should understand the distinction between working in an area of competence and working competently. Working in an area of competence requires members to operate within their qualifications and experience. Working competently requires sound judgement. If an error of judgement occurs the outcome may be construed as negligence; however, it does not necessarily imply that the member has acted unethically. Should consequent processes, including dispute resolution, reveal unethical behaviour the member concerned may subsequently face a further investigation under the disciplinary regulations.

Members:

a. Shall neither falsify nor misrepresent their own, or their associates' qualifications, experience and prior responsibility.

b. In the practice of consulting engineering, shall not describe themselves, nor permit themselves to be described, nor act as consulting engineers unless they are eligible to be corporate members and occupying a position of professional independence and are either prepared to design and supervise engineering work or act as unbiased and independent advisers on engineering matters.

c. Shall inform their employers or clients, and make appropriate recommendations on obtaining further advice, if an assignment requires qualifications and experience outside their fields of competence.

d. Shall acknowledge that the terms 'professional engineer', 'engineer' or 'member of the engineering profession' are used to describe only those persons eligible to be graduate or corporate members of the institution. Members who are not so eligible shall not indicate that they possess such qualifications.

e. Shall acknowledge that the term 'engineering technologist' is used by the institution to describe only those persons eligible to be affiliates of the institution. Members who are not so eligible shall not indicate that they possess such qualifications.

f. Shall acknowledge that the term 'engineering associate' is used by the institution to describe only those persons eligible to be associates of the institution. Members who are not so eligible shall not indicate that they possess such qualifications.

Clients and employers

Members have a responsibility to provide loyal service to their employer or client for whom they should apply their knowledge and skills with fairness, honesty and in good faith. Such loyalty extends to informing the employer or client of any possible adverse consequences of proposed activities based on accepted engineering practice of the day and taking all reasonable steps to find alternative solutions. Loyalty to the employer or client also requires that strict confidentiality be applied with respect to information or property available to the member as a result of the service provided. Members should not reveal facts, data or information obtained without the prior consent of its owner. The only exception to the provision of loyal service which can be condoned is when the welfare, health or safety of the community, or the environment on which they depend, is threatened by actions of the employer or client and all attempts to have the employer or client modify the proposed actions have been unsuccessful.

Members should relate to an employer or client on an open and informed basis so as to establish a position of trust. Any circumstances which may be regarded as detrimental to the maintenance of trust should be avoided or disclosed.

Members:

a. Shall promote the principle of selection of consulting engineers by clients upon the basis of merit, and shall not compete with other consulting engineers on the basis of fees alone. It shall not be a breach of the code of ethics for members to provide information as to the basis upon which they usually charge fees for particular types of work. Also, it shall not be a breach of the code of ethics for members to submit a proposal for the carrying out of work which proposal includes, in addition to a technical proposal, an indication of the resources which members can provide and information as to the basis upon which fees will be charged or as to the amount of the fees for the work which is proposed to be done. In this respect it is immaterial whether or not members are aware that others may have been requested to submit proposals, including fee proposals, for the same work.

b. May use circumspect advertising (which includes direct approaches to prospective clients by any reasonable means) which is not misleading, to announce their practice and availability. Information given must be truthful, factual and free from ostentatious or laudatory expressions or implications.

c. Shall, when acting as administrator of a contract, be impartial as between the parties in the interpretation of the contract. This requirement of impartiality shall not diminish the duty of members to fairly apply their skill and knowledge in the interests of their employers or clients.

d. Shall keep their employers or clients fully informed on all matters, including financial interests, which are likely to lead to a conflict of interest.

e. Shall advise their clients or employers when they judge that a project will not be viable, whether on the basis of commercial, technical, environmental or any other such risk which the member might reasonably have been expected to consider.

f. Shall inform their clients or employers of the possible consequences in the event that a member's judgements are overruled on matters relating to the welfare of the community. Where justified by the consequences which result from the matter continuing, members shall endeavour further to persuade the client or employer to discontinue with the matter. If unsuccessful, members may make the details of the adverse consequences known to the public without incurring a breach of the code of ethics.

g. Shall neither disclose nor use confidential information gained in the course of their employment without express permission, unless permission unduly withheld would jeopardise the welfare, health or safety of the community.

h. Shall not undertake, nor should they be expected to undertake, professional work without remuneration which is adequate to ensure that they are able to carry out their responsibilities in accordance with recognised professional standards.

i. Shall not accept compensation, financial or otherwise, from more than one party for services on the same project, nor provide free services, unless the circumstances are fully disclosed to, and agreed to, by all interested parties.

j. Shall neither solicit nor accept financial or other valuable considerations, including free engineering designs, from material or equipment suppliers for specifying their products.

k. Shall neither pay nor offer directly or indirectly inducements to secure work.

l. Shall neither solicit nor accept gratuities, directly or indirectly, from contractors, their agents, or other parties dealing with their clients or employers in connection with work for which they are responsible.

Colleagues

The tenets of the code of ethics are based on shared values and a shared responsibility to uphold them. Members have an obligation to exercise fairness in dealing with others and to provide support and assistance when required. Members should avoid any actions or statements which can be construed as being unfairly critical of a colleague or intended to favour their own position at the expense of a colleague.

Members:

a. Shall exercise due restraint in explaining their own work, shall give proper credit to those to whom proper credit is due and shall acknowledge the contributions of subordinates and others.

b. Shall accept, as well as give, honest and fair professional criticism when commenting on another's work or making public comment.

c. Shall compete on the basis of merit and not compete unfairly.

d. Shall neither maliciously nor carelessly do anything to injure, directly or indirectly, the reputation, prospects or business of others.

e. Shall, where acting as a representative on behalf of an employer, recognise that other members, who are employees, are colleagues to whom the code of ethics applies.

f. Shall uphold the principle of adequate and appropriate remuneration.

g. Shall neither attempt to supplant another individual or organization who has been duly appointed by a client or employer nor accept engagement from a client or employer in replacement of another without first ascertaining that the appointment has been terminated by due notice.

h. Shall examine the circumstances and determine the appropriateness of accepting an engagement from a client if they have evidence that they are to replace another, having first made all reasonable efforts to make the other aware of the situation.

i. Shall, if asked by a client to review the work of another, discuss the review with the other person or organisation prior to submitting the review if it is possible to do so.

j. Shall not unfairly criticise others for their past work where such work was conducted in accordance with the accepted standards and practices and community values of the time, and in accordance with the needs of the time.

k. Shall not continue in a business association with, nor practise with, any person who has been removed from membership of the institution because of unethical conduct.

Acting as an expert witness

An expert witness provides a special and unique service to legal or quasi-legal proceedings established for the purpose of making judgements. Once accepted by the judge or arbitrator, an expert witness is normally afforded two important privileges: the freedom to remain in the proceedings at all times and the freedom to express an opinion.

At all times the expert witness owes the proceedings total objectivity. The role of expert witness is to give the tribunal the benefit of his or her special training and experience in order to help the tribunal understand matters which it would not otherwise understand and thus help the tribunal to come to the right decision.

This duty to the tribunal is not inconsistent with the duty the expert owes to the client. In fact, the best way to discharge this duty is to be completely non-partisan.

An expert is not an advocate. Advocacy by an expert diminishes the value of advice both to the client and to the proceedings.

It follows that:

a. Members' reports, statements or testimony before any tribunal shall be objective and accurate. They shall express an opinion only on the basis of adequate knowledge and technical competence in the area, but this shall not preclude a considered speculation based intuitively on experience and wide relevant knowledge.

b. Members shall reveal the existence of any interest, pecuniary or otherwise, that could be taken to affect their judgement in a technical matter about which they are making a statement or giving evidence.

c. Members should ensure that all reports and opinions given to a client prior to a hearing include all relevant matters of which they are aware, whether they are favourable or unfavourable.

d. Members giving evidence as experts should listen very carefully to the question put, and ensure that each answer is given objectively, truthfully and completely and covers all matters relevant to the question of which they have knowledge.

e. When discharging these responsibilities, members should have regard to the normal practice at the time of the occurrence of the incident which gave rise to the call for advice.

Public comments or statements

Public comment and statements by members should comply with generally accepted standards of the community. The presentation of arguments should be made in a way that maintains and enhances community trust in the values and expertise of the membership of the institution. A loss of community trust would be contrary to the best interests of the community in circumstances where the members' comments might be crucial to the welfare, health and safety of the community. Members should display restraint in the manner in which they comment on engineering matters, especially in circumstances where the member, by explicit reference or implication, gives the public reason to believe that their comments are made on the basis of relevant knowledge.

It follows that:

a. Members may, if they consider that by so doing they can constructively advance the well-being of the community, contribute to public discussion on engineering matters in their area of competence.

b. In areas outside of a member's area of competence, but those in which a member can demonstrate adequate knowledge, comment may be made on details of a project within that area of knowledge. Adequate knowledge generally applies to a narrow aspect of an area of competence. Adequate knowledge may be acquired from working in a related area of competence or through continued professional development. However, adequate knowledge in a narrow area is not generally a sufficient basis for public comment or advice on the overall solution to an engineering task outside of a member's area of competence.

c. In areas outside of a member's area of competence, and in which the member is not able to demonstrate adequate knowledge, public comment or statements should be limited to enquiries which seek to provide deeper understanding. In this respect the member may draw on experience in engineering training and analysis as a basis for asking objective questions which may assist the public to evaluate engineering works without the member implying personal competence or knowledge in the area.

Whistleblowing

In the course of a member's employment, situations may arise concerning the employer or client organisation, which may present the member with a significant moral problem. These could include criminal behaviour, threats to public safety or unethical policies. The member has a responsibility under the code of ethics to ensure that any such practices are brought to the attention of those with direct authority to rectify the problem or, if the warnings are not acted upon, to raise the matter elsewhere. A decision to undertake an act of whistleblowing is a serious matter and the member must be aware of the personal costs that may be involved.

Whistleblowing differs from the broader aspects of public comment or statements in that it normally involves access to privileged information, either directly or indirectly, which is not otherwise in the public domain. Comment on the information available may lie outside a member's area of competence.

Because of the complex nature of the issue of whistleblowing the following practical and common sense guidance is set out for the benefit of members;

Make any objections to unethical practices promptly so as to avoid any misinterpretation of the motives for doing so.

Focus on the issues and proceed in a tactful, low-key manner to avoid unnecessary personal antagonism which might distract attention from solving the problem.

Keep supervisors informed of your actions, as much as possible, through both informal discussion and formal memoranda.

Be accurate in your observations and claims, and keep formal records documenting relevant events.

Raise the problem initially through normal organisational channels.

Consult colleagues for advice and avoid isolation.

Consult with the institution through the chief executive on the ethical issues involved, or with other organisations as appropriate.

Seek legal advice concerning potential legal liabilities.

Reference to material that may be of assistance to members is available from the institution's national office.

Scope of application and disciplinary procedures of the code of ethics

The code applies to all institution members, and to non-members who have agreed to be bound by them under any arrangement approved by council.

The membership

The provisions of the code are not limited by the geographic location of the member, except in any circumstance where their compliance would represent a breach of the laws or regulations of the location concerned. Collectively, the institution's membership comprises the following:

Professional Engineers: persons who have completed an engineering degree accredited by the institution or who have obtained other Australian or overseas qualifications and experience to a standard recognised by the institution as equivalent to such qualifications.

The institution adopts internationally recognised criteria for admission of such persons as professional engineers in the grade of graduate, and for advancement to the corporate membership grades.

Engineering Technologists: persons who have completed a course in engineering technology or other relevant disciplines, accredited by the institution or who have obtained other Australian or overseas qualifications and experience to a standard recognised by the institution as equivalent to such qualifications.

The institution establishes criteria for admission of such persons as engineering technologists in the grade of affiliate. This grade also includes those who have a three-year degree in a relevant science and an active interest in the engineering field.

Engineering Associates: persons who have completed a recognised Australian associate diploma and related work experience in a technical field of engineering, or who have obtained other Australian or overseas qualifications and experience to a recognised equivalent standard.

The institution establishes criteria for admission of such persons as engineering associates in the grade of associate.

National Professional Engineers Register (NPER)

The National Professional Engineers Register (NPER) is a register administered by the institution for professional engineers who meet stringent qualification, experience and continuing professional development criteria. When a professional engineer is entered in the register, he or she acknowledges a commitment to ethical practice and a willingness to maintain an appropriate level of professional competence through continuing professional development. Non-members of the institution may apply for registration.

Section 3 of the National Professional Engineers Register (NPER-3) is reserved for practising professional engineers. The register identifies the disciplines in which practitioners can demonstrate the skills, knowledge and experience appropriate for independent practice.

Procedures for handling alleged breaches of the code

The council of the institution has approved regulations to govern the investigation of alleged breaches of the code of ethics. The regulations provide for a process to investigate alleged breaches and to reflect the importance which the council places on all members upholding the ethical standards of the membership.

When a complaint is received by the institution an attempt is made to reach a resolution through conciliation. Subsequently, if necessary, the matter is examined by the chief executive to determine whether or not a formal investigation is required. In coming to a decision the chief executive considers the nature of the evidence submitted and whether or not the matters giving rise to the complaint, if substantiated, would amount to improper conduct. If a conciliator's report is available, this is also taken into account.

Complaints of a minor nature are decided by a senior office bearer. More serious complaints are referred to an investigating panel. A hearing may be held and, if necessary, witnesses will be called.

The following sanctions may be applied: admonition, reprimand, a fine, suspension of membership, deregistration, expulsion from the institution.

Details of the decision and the reasons for it are sent to the member concerned, who may lodge an appeal. The regulations provide for appeals to be heard by an appeals board.

Where breaches are proven, the decision is normally published. Where appropriate, similar publicity will also be given to complaints which are dismissed.

A booklet published by the institution, entitled *'Disciplinary Regulations and Regulations for Dealing with Failure to Maintain Appropriate Engineering Standards'* is obtainable free from any office of the institution on request.

Application to other professional engineering organisations

The Councils of the Institution of Engineers, Australia, the Association of Professional Engineers Scientists and Managers, Australia, and the Association of Consulting Engineers, Australia, have each adopted the provisions of this code as binding on the actions of members of their respective organisations. In this regard the councils have jointly advised and recommend to all professional engineers in Australia that the interests of the community and of their profession will be best served by commitment to the provisions of the code of ethics through full individual membership and active support of each of the organisations for which they are eligible.

Related institution policy statements and documents

- Environment Principles for Engineers
- Policy on Sustainability.
- Policy on Occupational Health and Safety
- Policy on Continuing Education

Appendix F

Institute of Electrical and Electronics Engineers (IEEE USA)

Code of ethics

We, the members of the IEEE, in recognition of the importance of our technologies in affecting the quality of life throughout the world, and in accepting a personal obligation to our profession, its members and the communities we serve, do hereby commit ourselves to the highest ethical and professional conduct and agree:

1. To accept responsibility in making engineering decisions consistent with the safety, health and welfare of the public, and to disclose promptly factors that might endanger the public or the environment.
2. To avoid real or perceived conflicts of interest whenever possible, and to disclose them to affected parties when they do exist.
3. To be honest and realistic in stating claims or estimates based on available data.
4. To reject bribery in all its forms.
5. To improve the understanding of technology, its appropriate application, and potential consequences.
6. To maintain and improve our technical competence and to undertake technological tasks for others only if qualified by training or experience, or after full disclosure of pertinent limitations.
7. To seek, accept, and offer honest criticism of technical work, to acknowledge and correct errors, and to credit properly the contributions of others.
8. To treat fairly all persons regardless of such factors as race, religion, gender, disability, age, or national origin.

9. To avoid injuring others, their property, reputation, or employment by false or malicious action.
10. To assist colleagues and co-workers in their professional development and to support them in following this code of ethics.

Approved by the IEEE Board of Directors, August 1990

Appendix G

The California Contractor's Code of Ethics (CC USA)

Center for Construction Education, establishes the Code of Ethics for Contractors.

Contractor code of ethics

- We will always provide a detailed cost breakdown before beginning work.
- We will always provide a written schedule for the progress of the work.
- We will always secure a building permit if one is required.
- We will always sign a written contract that includes our cost breakdown, the schedule, and the plans.
- We will always provide sufficient labor to accomplish the work according to the project schedule.
- We will always leave the job site in a clean and orderly condition.
- We will provide you with a list of the subcontractors that we intend to hire.
- We will provide a detailed invoice that incorporates a current report of the project budget.
- We will compile a booklet of all manufacturers' warranties for any new equipment we install.
- We will conduct a 'walk-through' inspection with you and correct any deficient work prior to final payment.
- Doing it right a contractor's standard for excellence.

5 *Ethics and business I*

A man was walking across the moor when a balloon appeared and hovered close by him. The balloonist leaned over and said to him, 'Could you possibly help me. I am aiming to meet a friend for lunch in this area but have got absolutely lost. Could you tell me where I am?' The walker looked at his map and after a long pause and several calculations replied, 'You are . . . degrees latitude and . . . degrees longitude.' The balloonist became agitated at this point, saying, 'You must be an engineer.' 'You're right,' came the reply 'but how did you know?' 'Well,' said the balloonist, 'You have given me a lot of information that I can't understand. I still have no idea about what I am going to do, and now you have made me even later.'

The walker replied, 'You must be a manager.' 'Yes', replied the balloonist, 'but how did you know that?'

'Well', said the engineer, 'you don't know where you are, where you are going or how you are going to get there. You have got to where you are through a great deal of hot air. And now that it has all gone wrong, somehow it's somebody else's fault.'

Introduction

Up to this point we have focused on professional ethics, reflecting on the purpose and responsibilities of the engineer. However, as we noted in the Challenger case, today's engineer does not always function simply in a consultancy role or in a purely professional role, as distinct from the role of manager or director. In this chapter we begin to explore the ethics of management. As the joke above suggests, it is very easy to see management in a stereotypical way – 'the people who don't really know about the technical details of engineering and who don't really know how to manage'. It is equally easy to see the perspective of business as something that conflicts with the profession of engineering. 'What has ethics got to do with business?' is the common cry. The extreme response to this can be found in the case of WorldCom.

Case 5.1

WorldCom was at one point the second largest telephone company in the United States. In 1997 they merged with MCI, costing some $37 billion (1 billion = 10^{12}). The success of the company hit problems in 2000, with the telecommunications industry as a whole suffering decline. Between 1999 and 2002 the company used fraudulent accounting methods to obscure its declining financial condition. It gave a false report of increasing profits in order to maintain the value of WorldCom's stock. It did this by reporting revenues that had been inflated by input from bogus corporate accounts and by recording some expenses as capital. An internal audit initially revealed a $3.8 billion dollar fraud. A later estimate was that the company's assets had been inflated by over $11 billion.

In one sense this case can be seen simply as a legal matter, with the executives breaking the law and eventually being punished. In another sense ethical meaning is at the centre of this. The core values of the company were about success in the marketplace, and it was clear that they felt any means could be used to achieve that. There were few points of reflection on practice. Even when the accountants (initially Arthur Anderson) were examining the books the full story was not revealed. Some would suggest that audits were compromised, due to a conflict of interest, with the accountants trying to ensure that they retained lucrative contracts and thus not wanting to dig too much below the surface. The very complexity of the organization did not help to make any of the fraud transparent. As we have noted in the Challenger case, large corporations lead to fragmentation of practice and responsibility, with few people actually being aware of all that is going on, and thus unable or unwilling to take responsibility for the whole. In such an atmosphere there is little responsibility and no sense that the main players could actually be called to account for their actions. Ultimately, of course, they were called to account, and the company and many stakeholders were ruined.

This case shows that whilst ethics in business is about individual behaviour it is also about the very organization and identity of the business itself. Like professional ethics, business ethics is at the heart of professional practice, and without clear understanding of responsibilities to stakeholders and clear accountability, any idea of purpose is lost. With a lost purpose and accountability, then, this inevitably affects the company's reputation and thus the basis of trust with consumers and the wider society. It is at its most extreme in firms such as WorldCom but is equally true of companies of any size.

In this chapter we will explore ethics in business and examine:

- The background to business ethics and why it is increasingly seen as important.
- Corporate social responsibility (CSR) and how this relates to business ethics.

- The grounds for developing CSR in business.
- An example of CSR policy.
- The subsequent and much broader view of ethics that emerges from CSR.

Business ethics

If ethics is the systematic study of right conduct then business ethics is ethics applied to the context of business. Business ethics includes exploration of:

- The underlying values of business, including those of any particular professions in business, such as accountants or managers.
- How values might be embodied in the corporation. This includes the development of codes of ethics (Webley 2003).
- Particular policies in areas such as corporate governance or workplace relationships.
- The wider responsibilities of business, to the local community and the environment and in global issues.
- Underlying ethical theories. The two most popular, as already described, have been that ethics should be based on core general principles (ontological theory) and that ethics should be based on maximizing the good for the greatest number (utilitarian theory). Increasingly, ethical theory is being based in the idea of character and the qualities that make up that character (virtue ethics).

For many companies business ethics and CSR are seen as one and the same. They involve various ethical problems that can arise in a business setting and any special duties or obligations that apply to persons who are engaged in commerce. While there are some exceptions, business ethics are usually less concerned with the foundations of ethics or with justifying the most basic ethical principles, and more concerned with practical problems and applications, and any specific duties that might apply to business relationships. This can be applicable at many levels to a number of situations within the construction industry such as bidding, tendering and contractual issues.

An ethical policy, code of conduct or a set of business principles can be used as a management tool for establishing and articulating the corporate values, responsibilities, obligations and ethical ambitions of an organization that dictates the way it functions. It provides guidance to employees on how to handle situations that pose a dilemma between alternative right courses of action, or when faced with pressure to consider right and wrong.[1]

The purpose of business ethics policies is to communicate a company's values and standards of ethical business conduct to employees and, beyond, to the community. It must also establish a companywide process to assist employees in obtaining guidance and resolving questions regarding compliance with the company's standards of conduct and values as well as establishing criteria for ethics education and awareness.[2]

It must ensure that all facets of a business will be conducted fairly, impartially, in an ethical and proper manner, in accordance with the values and code of conduct and in full compliance with all laws and regulations. In the course of business, integrity must underlie all relationships, including those with customers, suppliers and the community amongst employees. The highest standards of ethical business conduct and compliance is required of all employees in the performance of company responsibilities.[3]

> WorlCom survived their bankruptcy. Imagine that you are the new board and that it is your task to work out a new ethical code for the company. Detail the ten most important things you would want to be included in that code.

The importance of business ethics

Ethics plays an important part in business today, and the public interest and accountability in this sector are continually increasing. In the 1960s and the 1970s business came under increased scrutiny in the areas of equal opportunities and health and safety at work. This led to the establishment of legal standards, which have been a continuing feature of the defining of business ethics (Reed 1999).

Moeller and Erdal (2003, 3–4) note several other elements in the experience of business that have led to the increased concern for business ethics:

- *Globalization.* Globalization has seen the growth of multinational business such that it is estimated that over a half of the biggest economies in the world are corporations (Zadek 2001). Where such companies could act with apparent impunity at one time, their power and the fact of having to relate closely to different governments worldwide has led to increased sense of accountability for their actions.
- *Information and communications technology (ICT).* This has led to increased global transparency, with instant access to immediate information and judgements from many different sources. The result is that companies find it less and less easy to hide what might be a more controversial aspect of their business.
- *Fiscal pressure.* Growing fiscal pressure has forced companies to pull out of previous philanthropic ventures. At the same time, this has led to greater discussion

about the different roles of government and business and how business can best contribute.

- *The growing importance of intangible values.* As Zadek (2001, 7) notes, this has involved a recognition that in the new economic environment there are an increasing number of values shared by significant parts of society on which depend the continued success of the corporation. This is partly about increased awareness in society of key issues such as the sustainability of the environment, and the need to respond.

In addition to these there has been an enormous growth in non-governmental organizations (NGOs), such Oxfam, Christian Aid or Amnesty International. The term NGO was coined in 1945 by the United Nations and refers to groups that are independent from government and that seek to pursue public or common interest agendas. Their work includes empowering people and groups in areas of poverty or oppression; service delivery, including humanitarian aid and lobbying powerful decision-makers in politics and business to take account of social issues. The last of these functions can bring an NGO directly into conflict with business, especially in the global context, and frequently raises issues regarding the responsibility of business. A good example of this is the Brent Spar incident (see below, Entine 2002). Non-governmental organizations are not always right and do not represent any particular constituency. They have, however, played an important part in raising issues of corporate responsibility, in the continued dialogue about what the CSR of business might be and in the development of a globally transparent society.

Higher demands are thus being placed on businesses, including engineering firms, to implement more ethical, ecological and sustainable business practices. More ethical or perceived ethical behaviour can be good for and enhance business.

In addition, high-profile 'ethical disasters' at the beginning of the twenty-first century have focused the public and business community's attention on the need for clear ethical standards and behaviour. Enron, WorldCom, Boeing, Parmalat and Shell were, until recently, symbols of successful economic growth and model companies setting the pace for business (Boddy 2005, Sims and Brinkmann 2003). Now, many of these company names have become synonymous with greed and corporate malfeasance, resulting in tremendous losses in shareholder value and trust and in legal battles.

Corporate social responsibility

Before the issue of CSR can be explored it is important to have a clear understanding of what a corporation is and of its key features. In a general sense, a corporation is a business entity that is given many of the same legal rights as an actual person in today's society. A corporation may be made up of a single person or a group of people, granting a limited protection to the people involved in its business. A corporation may issue stock, either private or public; if stock is issued, the corporation will usually be governed by its shareholders. The most common model is that of a board of directors who make all major decisions for the corporation, in theory serving the best interests of the individual shareholders.[4]

Now the issue of whether a corporation is morally responsible for its actions should be raised. If a corporation is a collection of people working under the same roof, are individuals responsible for the results of their actions? Or is the corporation a legal and moral entity that takes responsibility for actions and decisions made by individuals within it? This debate is long, ongoing and complex. However, the general consensus is that moral and social responsibility is allocated to the corporation, rather than the individuals that work within. This argument is based on the idea that in order to assign responsibility to a corporation it must have legal independence from its constituent members. This is based on the theory that organizations have internal decision structures that are put in place to ensure that corporate decisions fall within the lines of the organization's predetermined goals. Therefore, decision-making is the responsibility of a number of people in the companies' decision structure, and not solely of individuals within a corporation. This internal decision structure aids in the establishment of corporate policies and determines the company's actions beyond the contribution of individuals (Crane and Matten 2004).

Another argument for the designation of moral and social responsibility towards corporations rather than individuals is that in conjunction with the internal decision structures, corporations also have a set of values, beliefs and protocols that lay out what the company regards as right and wrong. This is known as the organization culture. The organization culture within a company has widespread influences on individuals' moral and ethical decision-making and behaviour; therefore the individual making decisions lies under the net of the corporation as a parent, and generally reflects its core beliefs in the decisions made (Crane and Matten 2004).

The International Business Leaders Forum (IBLF) define CSR as,

> [O]pen and transparent business practices that are based on ethical values and respect for employees, communities and the environment. It is designed to deliver sustainable value to society at large, as well as to shareholders.
>
> (http://www.iblf.org/csr/csrwebassist.nsf/content/a1.html)

At one time the stress was on the 'social', with the view that CSR was largely about how the company contributed towards the local and possibly wider community. This was seen as a philanthropic view. The twenty-first century has seen

CSR expanding to take in an awareness of and responsiveness to all aspects of business life in the social and physical environment and in the company itself, a view that stresses interconnection and interdependence. Hence, the grounds for CSR are tied closely to the definition.

The grounds for corporate social responsibility

We will first look at the underlying arguments about the nature and justification of CSR, and then sum up the reasons for being involved in it.

The liberal view of corporate social responsibility

For Milton Friedman (1983) the answer was simple. The role of business is the creation of wealth and thus the prime responsibility of business is to make a profit. The exclusive duty of the company is to its owners, usually the shareholders. In this the executive acts as 'an agent serving the interests of his principal' (1983, 240). The interest of the principal is profit maximization, and involvement in any activities in the community outside this sphere would be a violation of trust and thus morally wrong. Ultimately, any money that is being used by a company executive for social concerns is the shareholders, and therefore cannot be used other than for making profits for them. Friedman does not argue against the social involvement of the company as such, rather simply that the company, and the owners especially, can decide to do what they think is fit. There can be no moral or legal constraint on the company to be more socially aware and the question of the rights of the community or groups within the community taking precedence over the company, as long as it is pursuing legal ends.

If the company executive does decide to become involved in a community project, Freidman argues that this is not the payment of an obligation but rather a means of achieving the company's aims. Thus donations to a local medical project, for instance, are not fulfilling obligations but rather improving the image and reputation of the company and thus contributing to improving profits.

Friedman notes the negative consequences of pursing social responsibility. It would involve costs that would have to be passed on to the customer, possibly to the shareholder in reduced dividends, and to the employee in reduced wages. Not only is this unfair, it also constitutes a form of taxation without representation and is therefore undemocratic. Moreover, it is both unwise, because it invests too much power in the executive, and futile, because it is likely that the costs imposed by this approach will lead to a reduction in economic efficiency.

Finally, Friedman argues that the executive is not the best person to be involved in making decisions about social involvement. He or she is neither qualified nor mandated to pursue social goals. Without the skills and experience of social administrators it is difficult to see how the executive could understand the needs of the local area or begin to determine local priorities. Such a task is better suited to local government and social concern groups, whose roles and accountability are directly related to these tasks. For business then to enter this field would lead to a confusion of roles and a raising of false expectations.

Friedman's argument focuses then on the primacy of the goals of profit making (written into the contracts of the executive), the obligations of the executive within such a contract, the freedom of the executive to pursue company goals and the local government as the proper focus for social activity. This approach is a good example of contract ethics in that responsibility is determined by the freely entered contract.

Friedman's argument is taken further by Elaine Sternberg (2000). Sternberg argues from a strong theoretical ethics basis, that of the telos. Aristotle argues that the idea of the good arises from reflection on the telos, the underlying purpose or end. Sternberg suggests that applied to business, the telos of business is to make profit and to stay within the law. Hence, any attempt to go beyond the basic telos would be teleopathic – a wrong end.

Albert Z. Carr provides an even narrower and more severe view of CSR. He argues that the ethics of everyday encounter have no part to play in business. Carr sees the so-called ethics of business as analogous to poker, a game where participants can reasonably aim to deceive other players. All the participants know the rules of the game and accept that they apply only in that situation.

Carr writes,

> That most businessmen are not indifferent to ethics in their private lives, everyone will agree. My point is that in their office lives they cease to be private citizens; they become game players who must be guided by a different set of standards and the golden rule, for all its value as an ideal for society is simply not feasible as a guide for business. A good part of the time the businessman is trying to do unto others what he hopes other will not do unto him.
>
> (Carr 1968)

On this basis, business people can acceptably deceive those in the business world, including the customer, with respect to advertising. All expect to be deceived and operate accordingly. So social responsibility takes second place to the game and the purpose and rules of the game.

All three arguments are based on what Berlin (1969) calls a negative view of freedom, i.e. freedom from oppression or interference. Hence, they seek to protect the agent and the shareholder from attempts to determine what they should do. They also rest on the issue of the purpose of business, and a narrow of view of that purpose is defended. Finally, they rest on the freedom to articulate in the marketplace, which for some is seen as the place where individual effort and effective distribution of wealth come together.

Michael Novak develops this in a different way. He suggests that the freedom to do one's duty is central to the argument, not just negative freedom, and that duty of the business person is of a high moral nature. The very activity of enterprise is a moral good. Novak (1990, 12) argues firstly that every human being has the right to personal economic activity. Indeed, the capacity for enterprise is a fundamental virtue that Novak infers is basic to human development and fulfilment. Secondly, in this context the business person is serving society and wealth creation is a socially responsible activity, both in the practice of his or her skill and in the distribution of wealth. Thirdly, underlying this, Novak links the moral perspective of the marketplace. The market actually enables not just the distribution of wealth but also the development of community. 'Markets', argues Novak, 'draw people out of isolation into reasoned, civil voluntary interchange with their fellows'. Markets, in other words, enable community. It is possible to argue from this that contracts enable empowerment and freedom. Individuals enter the contract freely, and operating in the market empowers them.

Novak then sees the marketplace as ethical and responsible in itself, a basis for the development of autonomy, the distribution of wealth and for building community. In such a context there is little need for a policy on CSR, because the business person is directly enabling society as a whole to grow financially and as a community.

The argument set out by these writers is not simply pragmatic but has a variety of philosophical underpinnings, involving underlying values, views of society and different ideas about purpose.

> Is there one core purpose of business? If so, how does that relate to any other role of business?
>
> Make a list of all the possible purposes of your business or university. How might a public organization, such as a university, differ from a business?

Critique

With respect to the view of society, Novak and Friedman have an idealized view of the market. In Novak's case this assumes an underlying equality, ignoring the power imbalances within the market that militate against simplistic community building, especially seen at a global level. All ignore the complexity of the relationship between the marketplace and society, and the subsequent responsibilities that emerge. Hence, whist the market is an important mechanism, more and more commentators prefer to look to ways in which the market can work with society (Buchholz and Rosenthal 2002).

Secondly, two key underlying values are freedom and creativity. Freedom is important but, as Tawney (1930) notes, it requires the balance of other values. The freedom of any person affects the freedom of another. The negative view of freedom tends to assume an individualistic view of the person. Against this is a view of humanity as interdependent. This suggests neither a collectivist view of society nor an individualist view. Creativity in that view is very much about partnership and about working together with the community.

Thirdly, Sternberg's view of the purpose of business is simplistic. It assumes a single purpose in relation to the immediate task of the businessperson. However, there is no reason why the business person may not have several different purposes each of equal importance, care for shareholders, clients, the physical environment and so on. Perhaps the biggest problem with the arguments from purpose is the assumption that shareholders have a single purpose. It is certainly true that one aim of shareholders is to make a profit. However, it is simply not clear that shareholders do not have an equal concern for the environment or for the community in which they live. This can only be tested in dialogue with each group of shareholders, and in the light of the nature of the business and its effects on society.

Fourthly, behind much of this is another assumption, that the different ethical worlds of social concern and business are discrete, quite separate. This is taken further by Carr, who would have us suspend the personal moral world in coming to work. In fact, the different value worlds are many and varied and are connected. It is not possible to simply say that there is a clearly defined wall between them. The shareholder, for example, is also a member of the community, with concerns for the community. Equally, as a business works in other countries it encounters a great many other cultural and religious values that affect different stakeholders and that in turn will affect how the company does business in that context. In this sense the company has to be aware of the plurality of values and be prepared to negotiate and work with them.

Fifthly, in this light it becomes difficult to predetermine what the responsibility of the businessperson or the business should be anymore than it is possible to be precise about the responsibility of, for instance, local or national government. In practice there are broad responsibilities but these are continuously being debated and negotiated. An example of this in government terms is the issue of respect. The responsibility for identifying and developing respect has traditionally been seen to be that of the school or the family. At the beginning of 2005 the UK government also began to argue that it had an important part to play in encouraging the definition and practice of respect (http://www.labour.org.uk/respect). The issue of respect is one that all groups in society have a concern for, including business.

Finally, if the shareholders and stakeholders are in creative dialogue it is possible to develop social enterprise that is good for all concerned. A good example here is the community involvement of Leeds Metropolitan University. Whilst higher education is based on public funds it is nonetheless increasingly a business that has to break even and that relies on developing the education market to achieve this.

Case 5.2
Leeds Met University

Leeds Met University set up a series of high-profile sponsorships in 2005, including Leeds Rhino's Rugby League Club, Leeds Tykes Rugby Union Club, Yorkshire County Cricket Club and the Black Dyke Mills Brass Band. This culminated in naming the famous Headingley sports ground (home of the three teams) as Headingley Carnegie (the sports faculty of the Leeds Met).

There was initial concern in the higher education world precisely because it was public money that was being used for this. In this sense the government could be seen as the chief shareholder.

The action, however, made good community sense and good business sense. Through the sponsorship, Leeds Met recognized its part as a member of the community and looked to support groups that were a key part of the culture of the community. This was good in itself (see corporate citizenship in Chapter 6). However, through this support, Leeds Met raised its profile immeasurably in the whole region and the country, and established a clear sense of its identity. An International Cricket Test Match, for instance, was played at Headingley Carnegie for over five days. This allied the Leeds Met identity to international sport and to the value world of cricket. In turn, this raised the profile in the community, in schools and with regional businesses, which began to see that Leeds Met was a good brand to associate with.

Such links in turn led to some remarkable creative projects. The links to the Black Dyke Mills Band produced important contributions to the culture of the university. It also led to the university and the band jointly developing the West Yorkshire Youth Brass Band. This in turn gave Leeds Met a higher profile in schools and colleges, especially with those who had a musical interest.

The Leeds Met case shows that CSR combines self and community interest, suggesting that shareholders and stakeholders should not be seen in terms of strictly separate interests.

Questions

How do your personal values relate to business values? How do your professional values relate to broader social values, such as equality or freedom?

How do the values of business relate to those of the community?

Stakeholder theory, ethics and social responsibility

Traditionally, so-called stakeholder theory (SHT) has been used as the basis for arguing against the liberal view of CSR. A stakeholder was initially defined in terms of those groups that were critical to the survival of the business, including employees, customers, lenders and suppliers (Sternberg 49). This concept has been further widened to include 'any individual or group who can affect or is affected by the actions, decisions, policies, practices or goals of the organization' (Carroll and Buchholtz 1996, B. 315). This results in a much more complex situation, encompassing the government, community and beyond. For multinational corporations (MNCs) this takes a particularly complex turn.

Heath and Norman (2004) suggest there are several different SHTs within this. They include:

- Strategic SHT, a theory that attention to the needs of stakeholders will lead to better outcomes for the business.
- SHT of governance, a theory of how shareholder groups should be involved in oversight of management, e.g. placing shareholders on the board.
- Deontic SHT, a theory that analyses the legitimate rights and needs of the different stakeholders and uses these data to develop company policies.

The debate continues as to whether there are really many different theories or whether they are parts of one major view of how business relates to society.

Sternberg (2000, 49 ff.) argues against the forms of theory that assert the rights of the stakeholder on several grounds:

- The argument that stakeholders should have equal representation on the board mistakenly confuses business with government.
- SHT rests on a confusion about accountability. Whilst business should take many groups into account it is not accountable to them. Sternberg's argument rests on the meaning of being called to account. If it is used for all stakeholders then the meaning of it becomes unclear.
- Faced by the many competing interests of the stakeholders there are no criteria offered by the theory for deciding how to handle conflicting interests.

One further danger of the SHT used in relation to CSR is that it can simply mirror the difficulties of the liberal position. Just as it is not possible to predetermine what the responsibility of the businessperson should be, it is also not possible to predetermine that business should fulfil the needs of all its stakeholders. Too easily, both positions move into rights arguments – the right of the executive to get on with his or her job and the right of the stakeholder to have his or her needs met.

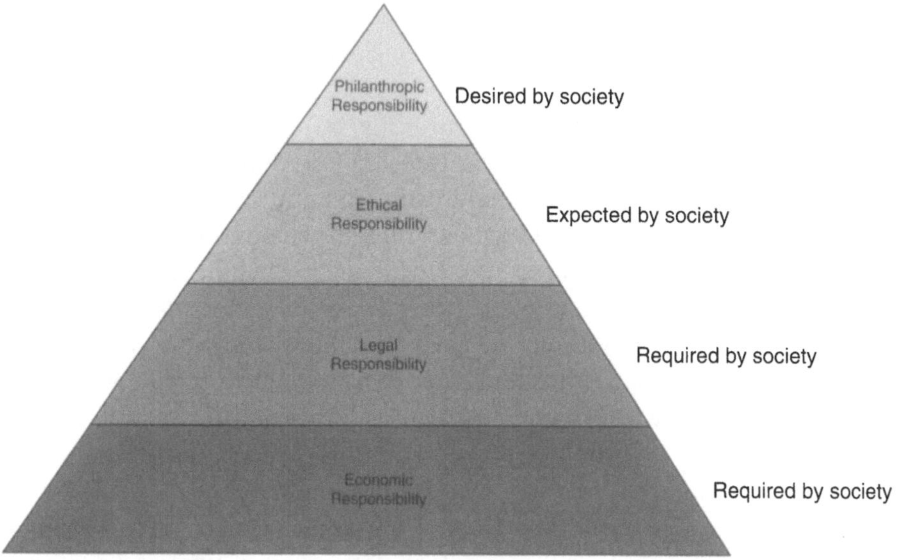

Figure 5.1: Carroll's four-part model of corporate social responsibility. Source: Carroll (1991) The pyramid of corporate social responsibility; towards the moral management of organisational stakeholders.

Carroll suggests a way of looking at CSR that gets over this polarized approach by suggesting that CSR has four areas (Figure 5.1): economic, legal, ethical and philanthropic. Carroll argues that these different responsibilities are set in consecutive layers within a company, with CSR involving the addressing of all the four layers consecutively.

Corporations have an economic responsibility towards their shareholders to be profitable, and provide reasonable returns on shareholders' investments. Economic and financial gain is the primary objective of a corporation in a business sense and is the foundation upon which all the other responsibilities rest.

However, at the same time businesses are expected to comply with the laws and regulations as the ground rules and legal framework under which they must operate. A company's legal responsibilities are seen as coexisting with economic responsibilities as fundamental precepts of the free enterprise system.

Ethical responsibilities within a corporation ensure that the organization performs in a manner consistent with expectations of ethics in society. It is important to recognize and respect new or evolving ethical trends adopted by society and that good corporate citizenship be defined as doing what is expected morally and ethically. It must be noted that the corporate integrity and ethical behaviour of a company go beyond mere compliance with laws and regulations and is the obligation to do what is right and fair, and to avoid harm.

Philanthropic responsibilities include corporate actions that are in response to society's expectation that businesses be good corporate citizens and engage in activities or programmes to promote human welfare or goodwill. Philanthropy, 'love of fellow human', is highly desired by society, argues Carroll; however, it is not ultimately necessary.

Carroll's view is comprehensive and usefully brings together different views of CSR. However, the distinction between desirability and necessity in the fourth area is not clear. It simply assumes that business is not primarily concerned with the promotion of human welfare. However, human welfare may at any point be a critical issue in the life of a business. If a subsidiary is directly abusing the human rights of its workers, for instance, then human welfare is not simply desirable, but is a moral demand. How one answers this moral demand will depend on the situation. Carroll's contention that CSR will operate in a manner consistent with the ethics of the wider society is also questionable. Does this mean that business should accept without question the ethics of a fascist society? A less extreme example might be a society that is undergoing transition, and with no fixed ethical expectation. As we shall note in the next chapters there are increasing examples of business contributing towards the ethics of wider society rather than following it.

As Buchholz and Rosenthal (2002, 316) note, feminist ethics, in particular, argue for a web of stakeholder relations that stress connectivity, interdependence, power sharing, collective action and conflict resolution. At one level this points to an empirical and developmental truth. Business is a part of society and its identity is established through how it relates to society, not least in its conduct with those who are affected by it, and who affect them. Hence, the identity of the company is continually developing, and has to be worked on.

This is further stressed by Heath and Norman (2004). They note how the view of shareholders' interests as opposing stakeholders' is false. Learning lessons from disasters such as the Enron case reveal that the real problems emerge from problematic managers who keep their actions secret from the shareholders. Hence, the shareholders were not able to be part of an adult conversation about values, purposes and practice of the business. The message is this: when business, or any corporation, is not transparent, or also fragmented, then the sense of responsibility of members, or the corporation as a whole, is easily lost.

Openly adopting high standards of CSR makes a positive contribution to society and can be regarded as a long-term business investment. It benefits the corporation by creating a more stable and improved context in which to do business. Corporate social responsibility is a tool that makes organizations act in a socially responsible manner. It can lead some companies to use this as a way of promoting the maximization of profits (Crane and Matten 2004). However, self-interest does not exclude social responsibility or vice versa.

It must be noted that all corporate activities and decisions have social impacts associated with them. These impacts can have positive and negative implications associated with them, and normally surface when providing products, services, employment and other corporate activities or business transactions (Crane and Matten 2004).

There are some industries that, by definition, could not develop any CSR policy. Is it possible, for instance, for an arms manufacturer or the tobacco industry to articulate a CSR policy? Directly or indirectly, both industries lead to immense suffering, which would make it impossible to square with a CSR policy. We are intentionally not making any judgements about these areas but rather invite you to use the ethical decision-making process outlined in Chapter 3 to look at the case for these areas and decide for yourself.

Questions

Imagine that you are on the board of a tobacco firm and that you have been charged with developing a CSR policy. How would it look and how would you justify it? (Palazzo and Richter 2005).

Imagine that you are on the board of an arms firm, with contracts for long-range weapons. How would you develop a CSR policy for this firm and how would you justify the purpose of the firm?

Compare the two policies and their justifications. Is one more acceptable than the other, if so in what way?

Watch the film *Thank You for Smoking*. What are the arguments set out for the tobacco industry there?

A good example of a CSR policy statement comes from the insurance group Aviva. Involving many different companies, Aviva sets out a statement that aims to bring together eight core policies around CSR.

AVIVA Group Corporate Social Responsibility Policy

Aviva as a member of the international community recognises its corporate social responsibility commitments in its various roles, which include insurer, investor, employer and consumer.

We reflect these commitments in a family of eight policies, which relate to:

Standards of business conduct
We are committed to ensuring that our business is conducted in all respects according to rigorous ethical, professional and legal standards.

Customers
We seek to provide our customers with a service hallmarked by integrity, quality and care.

Human rights
We respect the Universal Declaration of Human Rights and seek to be guided by its provisions in the conduct of our business.

Workforce
We are guided by our aim to be the employer of choice in all countries in which we operate.

Health and safety
We are committed to providing a working environment which is both safe and fit for the intended purpose and ensures that health and safety issues are a priority for all business operations.

Suppliers
We regard suppliers as our partners and work with them to help us to achieve our policy aspirations in the delivery of our products and services.

Community
We strive to be a good corporate citizen around the world, recognising our responsibility to work in partnership with the communities in which we operate.

Environment
We are committed to a programme of management, continuous improvement and reporting of our direct and indirect impacts, which marks our contribution to the world in which we live.

We recognise that our business activities have direct and indirect impacts on the societies in which we operate. We endeavour to manage these in a responsible manner, believing that sound and demonstrable performance in relation to corporate social responsibility policies and practices is a fundamental part of business success.

We are committed to continuous improvement in our corporate social responsibility programme and encourage our business partners to strive for matching performance.

Our business units throughout the world are committed to achievement of our policy objectives. Our performance will be periodically reviewed and externally verified to help us meet our policy goals. At a local level, the heads of our business will revise progress at least annually. We will publish reports regularly.

Approved by the Board of Aviva plc, January 2005
The Aviva CSR policy is framed in the context of basic ethical meaning that is summed up in terms of business standards. It seeks to identify core values and look to embed those values in practice.

Standards of business conduct

We regard ethical practice as critical to responsible business and we are committed to conducting all aspects of our business according to rigorous ethical, professional and legal standards
The spirit and guidance of the standards of business conduct policy embody our corporate values of *integrity, performance, progressiveness* and *teamwork.* The policy must be followed by all our staff, in particular by our senior managers across our businesses, to set an example for others to follow. All CEOs around the group must provide annual assurance that the policy is adhered to within their business. Compliance with the standards is also required as part of Aviva's senior management contract.

Measures to embed the standards of business conduct policy. Communicating to staff at the start of their career with us
Aviva's head office and Norwich Union Life communicate the Aviva Values and the standards of business conduct policy, respectively, to staff via their employment contracts. In Romania, where Aviva's reputation is considered to be one of its most precious assets, new insurance agents receive a copy of *The Agent Code* and employees receive the staff handbook. These explain that Aviva is in business for the long term and that we have values that shall be respected.

Communicating to staff during their careers
Many of our businesses around the group have translated the standards of business conduct policy into the local language (where appropriate), have tailored it to local needs and have made it available on the local intranet. The Leadership in Action course for all Aviva head office employees with high management potential continues to be successful in raising and embedding the Aviva Values through good leadership, focusing specifically on integrity, honesty and trust. The reinforcement of the Aviva Values takes place via employee awards in Aviva India and via employee meetings in our Spanish businesses.

Customer-facing ethical behaviour
Sales agent training is a focus for many of our European emerging market businesses, including those in the Czech Republic, Hungary, Romania, Poland and Lithuania. In Poland, the remuneration of sales agents is not only based on the amount of premiums bought but also on the quality of service offered to our customers.

Norwich Union Insurance's Leadership and Care programme reinforces the guidance of the Financial Services Authority's Treating Customers Fairly publications.

Monitoring our standards of business conduct

Performance evaluations
Behaviour in accordance with the Aviva Values is a core component of employee performance evaluations in Aviva's head office, Delta Lloyd Group in the Netherlands, Norwich Union Life and Hibernian in Ireland.

Electronic monitoring
All Aviva Canada employees are required to acknowledge online receipt, understanding and acceptance of the code of conduct, privacy policy and employee personal information policy annually. A similar annual electronic tracking system is in place in Aviva USA.

Maintaining registers
In the UK, Norwich Union Professional Services maintains a hospitality register and a register of interest. The registers have also been shared with other Aviva businesses, as an example of good practice.

Whistleblowing
While the majority of requests for investigations come through the line management, a 24-hour independently run telephone fraud reporting service also operates in all UK business units and a number of overseas businesses, including our offshore operations.

Work continues to expand this service to more businesses outside the UK.

Action against financial crime
In 2005, Norwich Union Insurance published a Fraud Report which revealed that fraud now costs the UK £16 billion a year and that it has grown by 15% in the past five years alone. To meet this challenge, we have implemented policies that meet the requirements of applicable legislation, regulatory guidance and industry good practice regarding fraud management, anti-money laundering and malpractice reporting. We have also introduced a single financial crime network across the group, bringing together expertise on fraud and anti-money laundering. Corporate social responsibility report 2006.

Customers are the life-blood of our business. We consider who they are or who they might be. We ask for their opinions. We listen to their views. We investigate their current and future needs. In the light of this continuing research, we modify our products and services. We provide some examples of the progress our businesses have made in embedding CSR in products and services in our three core business activities.

Aviva's approach involves several elements:

- A holistic approach that connects business ethics to CSR, and thus develops congruence between external and internal responsibilities.
- The inclusion of stakeholders, and in particular, customers, in dialogue and planning.
- A transparent framework that includes monitoring and regular reporting.
- Staff development that seeks to communicate core standards across the group. This is particularly important in light of the multinational nature of the group.

Hence, the old divide between business ethics, seen as the ethics of what goes on in the company and in relationships with customers, and CSR, seen as responsibility to society as a whole, is bridged, by dialogue, consistent practice and shared values.

Summary and conclusions

This chapter has shown that it is important for any business to be open and to attend to the place of business in society. This demands a close working relationship with the different stakeholders, and setting values and standards in policy and practice. Given the impetus for CSR and ethics, the whole area has moved on and points to a very different kind of business ethics. The old ethics was about something that was quite distinct from business decision-making with areas of social responsibility seen as philanthropy. The new ethics sees CSR as part of the everyday decision-making of business. It is about awareness of the social and physical environment and working out appropriate responses. In all this, codes of ethics are important guides, but are only useful as far as they reinforce the responsibility of the professional and the business as a whole. Equally important are the culture and ethos of the corporation and the capacity of the professional. These are critical if values and practice are to remain constant under pressure. Such values and awareness erode in the absence of a real learning organization such that values are embedded and practice monitored. Such monitoring will be examined more closely in subsequent chapters, especially around environmental ethics.

Cadbury (2002, 18) notes a good example, taken from Brooks (1964) of how easily companies can have excellent codes, but move into practice that is the opposite of the code. This was taken from the American company General Electric in the 1930s, but the story is remarkably 'modern'.

Case 5.3
General Electric

For the past eight years or so, General Electric has had a company rule called Directive Policy 20.5, which reads in part 'no employee shall enter into any understanding, agreement, plan or scheme, expressed or implied, formal or informal with any competitor with regard to prices.' The trouble, at least during the period covered by the court action and apparently for a long time before, was that some people at General Electric, including some of those who regularly signed 20.5, simply did not believe it was to be taken seriously. They assumed that it was window dressing, that it was in the book solely to provide legal protection for the company and for the higher-ups and that meeting illegally with competitors was recognized and accepted as standard practice and that often when a ranking executive ordered a subordinate executive to comply with 20.5 he was actually ordering him to violate it. Illogical as it might seem this last assumption becomes comprehensible in light of the fact that for a time when some executives orally conveyed or re-conveyed the order they were apparently in the habit of accompanying it with an unmistakable wink. . . . Asked, at a later inquiry, by Senator Kafauver how long he had been aware that orders issued in General Electric were sometimes accompanied by winks, Robert Paxton, a senior executive, replied that he had first observed the practice way back when his boss had given him an instruction along with a wink or its equivalent. [It was] sometime later that the significance of the gesture dawned on him, and he became so incensed that he had, with difficulty, restrained himself from jeopardizing his career by punching the boss in the nose. Paxton went on to say that his objection to the practice of winking had been so strong as to earn him a reputation in the company for being an anti-wink man. But he for his part had never winked.

Humorous though the case may be, it led to several senior managers going to jail, simply because they did not act with integrity, and because there was no monitoring of practice or questioning of behaviour.

Previous chapters showed how the business and professional spheres can conflict. This was clearly described in the Challenger case in Chapter 1. However, what this chapter shows is that professional and business ethics are not radically different as such. Business does produces some pressures that can pull the practitioner away from key values. However, as we have seen, there is real pressure, through concern for CSR, for business to articulate and maintain core values, and to articulate and fulfil responsibilities to stakeholders.

Questions

Go to the website of six major engineering industries. Do they have CSR policies? Map out the values that they articulate?

As the CEO of a major engineering firm what would your CSR policy be and what values would you set out in any mission statement?

References

Berlin, I. (1969). *Four Essays on Liberty*. Oxford: Oxford University Press.

Boddy, D. (2005). *Management: An Introduction*. Harlow: Prentice-Hall.

Brooks, J. (1964). *The Fate of the Edsel and Other Adventures*. London: Victor Gollancz.

Buchholz, R. and Rosenthal, S. (2002). Social responsibility and business ethics. In *A Companion to Business Ethics* (R. Frederick, ed.) pp. 303–22, Oxford: Blackwell.

Cadbury, A. (2002). Business dilemmas: ethical decision making in business. In *Case Histories in Business Ethics* (C. Megone and S. Robinson, eds.) pp. 9–22, London: Routledge.

Carr, A. (1968). Is business bluffing ethical?, *Harvard Business Review* 46 (January/February).

Carroll, A.B. and Buchholtz, A.K. (2000). *Business and Society – Ethics and Stakeholder Management*. London: Thompson.

Crane, A., and Matten, D. (2004). *Business Ethics – A European perspective – Managing Corporate Citizenship and Sustainability in the Age of Globalisation*. Oxford: Oxford University Press

Entine, J. (2002). Shell, Greenpeace and Brent Spar: the politics of dialogue. In *Case Histories in Business Ethics* (C. Megone and S. Robinson, eds) pp. 59–95, London: Routledge.

Friedman, M. (1983). The social responsibility of business is to increase its profits. In *Ethical Issues in Business* (T. Donaldson and P. Werhane, eds) pp. 239–43, New York: Prentice-Hall.

Heath, J. and Norman, W. (2004). Stakeholder theory, corporate governance and public management: what can the history of state-run enterprises teach us in the post-Enron era? *Journal of Business Ethics* **53**(3), 247–65.

Moeller, K. and Erdal, T. (2003). *Corporate Responsibility towards Society: A Local Perspective*. Brussels: European Foundation for the Improvement of Living and Working Conditions.

Novak, M. (1990). *Morality, Capitalism and Democracy*. London: IEA Health and Welfare Unit.

Palazzo, G. and Richter, U. (2005). CSR business as usual? The case of the tobacco industry. *Journal of Business Ethics*, **6**(4), 387–401.

Reed, D. (1999). Three realms of corporate responsibility: Distinguishing legitimacy, morality and ethics. *Journal of Business Ethics* **21**(1), 23–36.

Sims, R.R. and Brinkmann, J. (2003). Enron ethics (or culture matters more than codes). *Journal of Business Ethics*, **45**(3), 243–56.

Sternberg, E. (2000). *Just Business*. Oxford: Oxford University Press.

Tawney, R.H. (1930). *Equality*. London: Allen and Unwin.

Webley, S. (2003). *Developing a Code of Business Ethics*. London: Institute of Business Ethics.

Zadek, S. (2001). *The Civil Corporation: The New Economy of Corporate Citizenship*. Earthscan Publications Ltd, London.

Notes

1 http://www.ibe.org.uk/codesofconduct.html (accessed on 10 February 2006).

2 http://www.boeing.com/companyoffices/aboutus/ethics/(accessed on 8 February 2006).

3 http://www.boeing.com/companyoffices/aboutus/ethics/(accessed on 8 February 2006).

4 http://www.wisegeek.com/what-is-a-corporation.htm?referrer=adwords_campaign=corporation_ad=015471&_search_kw=what%20is%20a%20corporation (accessed on 13 February 2006).

6 *Ethics and business II*

While many definitions exist, as noted in Chapter 5, CSR looks at the socially responsible way companies relate to their stakeholders, across a whole dimension of activities. The essence is that companies behave in a responsible manner beyond their commercial and regulatory requirements. The CSR concept is closely allied to the emergence of the concept of 'corporate citizenship', which Andriof and McIntosh (2001) described as 'understanding and managing a company's wider influences on society for the benefit of the company and society as a whole'. The corporate citizenship approach takes the view that companies, as independent legal entities, are members of society and as such can be regarded as citizens with legal rights and duties. Accompanying these rights and duties is an expectation to act as good corporate citizens and contribute to the well-being of society more generally. While CSR is seen as an external-facing concept focusing on external relations, corporate citizenship requires the creation of internal structure perspectives before dealing with external relationships. Good corporate citizenship is a part of every facet of the organization.

Corporate social responsibility is also linked to other concepts such as social cohesion and social capital. Conceptually, CSR is primarily associated with voluntary integration of social and environmental concerns in organizational strategy. However, there is a growing recognition of the importance of business in developing social cohesion and helping overcome societal problems. The stakeholder concept is critical to the role organizations play in developing social capital. Dialogue with various stakeholders helps develop this social capital. There are many definitions of social capital and for the purposes of this chapter we will use that set out by the Report on Social Capital and Economic Development in the northeast of England.

Social capital is defined as the personal contacts and social networks that generate shared understandings, trust and reciprocity within and between social groups, and which underpin cooperative and collective action, the basis of economic prosperity and economic inclusion.

The company in the light of corporate citizenship can see itself as a member of society, rather than a discrete organization. Social capital is developed not by one-off gifts to the community but by commitment of the company to projects that make a difference to that community. The resources given are thus part of an ongoing relationship. Good examples of this are involvement with local schools or colleges, contributing to curriculum development or widening participation. One company, for instance, has contributed to the development of a local peace museum, with the explicit condition that the museum should develop a programme for conflict resolution in local schools. Such a company can then find itself setting part of the local ethical agenda and enabling ethical education in partnership with others.

This chapter focuses on the development of CSR and corporate citizenship and in particular the issues that provide a framework for good corporate behaviour. These include:

- CSR guideline and codes
- Accountability and corporate governance
- Business conduct
- Bribery and corruption
- Political issues
- Whistleblowing
- Supply chain management
- Community involvement
- Labour and human rights
- Health and safety

Corporate social responsibility – guidelines and codes

Corporate social responsibility by its very definition is open to interpretation when it goes beyond the regulatory and legislative frameworks. What then constitutes responsible corporate behaviour for an engineering business? A number of guidelines and codes of conduct have been created to provide guidance and allow scope for comparison in behaviour. These guidelines include:

- The Global Sullivan Principles
- The UN Global Compact
- SA 8000

- Caux Principles
- Global Reporting Initiative
- OECD Guidelines for Multi-Nationals
- The Keidanren Charter for Good Corporate Behaviour
- Principles for Global Corporate Responsibility

Each of these approaches, several of which will be examined more closely in ensuing chapters, provides a basis for companies to structure and demonstrate their corporate behaviour. Certain themes that are common to them all are summed up in the following headings.

Accountability and corporate governance

A central pillar of developing a CSR philosophy is the willingness of a firm to be accountable for its behaviour. Accountability in the ethical corporation goes beyond operating within the law and regulatory framework, to being responsible for the organization's actions and its impacts. The social, environmental and economic footprint of an organization is relevant to this accountability. Solomon and Solomon (2004) suggest that 'corporate governance is a system of checks and balances, both internal and external to companies, to ensure that they discharge accountability to all their stakeholders and act in a socially responsible way in all areas of their business'. This definition is based on the perception that companies can maximize value creation over the long term, by discharging their accountability to all their stakeholders and by optimizing their system of corporate governance.

Recent problems with corporate governance have resulted in a series of regulatory guidance and codes of practice. In the United Kingdom, the Cadbury (1992), Turnbull (2005) and Higgs (2003) reports have all resulted in a strengthening of the combined codes of practices for corporate governance. The Sabannes Oxley Act in the United States attempts to exercise similar controls following the collapse of Enron. All the initiatives relate to the manner in which the company is controlled. In the United Kingdom, the Company Law Reform Bill, 2006, is attempting to further tighten up the regulatory climate for companies in the manner of the Sabannes Oxley Act. Ultimately, governance relates to the integrity and ethical behaviour of those that lead the organization. The principles of corporate governance include the following:

Transparency. It is commonly accepted that transparency is an essential element of a good system of corporate governance. Internal control and the audit function are necessary to ensure transparency. The failings at Enron and Worldcom can be traced to the lack of transparency and the failure of the audit system. Directly associated with transparency is internal control. This relates to the system of controls for all aspects of the business, although traditionally internal control was limited to financial and operational aspects of the business. Internal control working effectively can reduce the level of risk facing a company. These risks may be financial, environmental, social or legal. Improved risk management reduces the prospect of negative social performance.

Disclosure. All companies produce a substantial amount of regulatory information such as annual reports and other regulatory items. Many other communications such as press releases, publicity material and websites are also produced. The lack of a structured system of disclosure would make it difficult for stakeholders to assess the accuracy and reliability of this information. The traditional audit process was supposed to check the veracity of financial reporting. However, Enron and subsequent revelations in financial reporting systems suggests that financial disclosure has not been to the standards expected in the underlying stewardship principles. Disclosure can work as a positive tool for better corporate performance. Increasingly, companies are adopting the use of combined safety, health and environmental reports or corporate responsibility report to disclose performance in these sensitive areas.

Leadership. The development of accountability in a firm is driven by senior management and is ultimately a question of leadership. The personal ethical ideals and integrity of these leaders determines the stance the organization takes on ethical issues. In larger, publicly accountable firms a code of conduct is set out that helps promote ethical and responsible decision-making. Where there is a greater accountability to shareholders, the senior management board should be in a position to challenge and review management performance. Agency theory indicates that professional managers should act in the best interests of their shareholders while stakeholder theory suggests that the firm has accountability to a wide range of stakeholders. While convergence of these views would create an ideal situation, leaders of organizations are accountable for the actions of their organizations.

Business conduct

A central tenet of ethical behaviour is compliance with the law of the country in the execution of the activities of the company. The laws of a country are designed for the benefit of society. There is a plethora of laws with which a company must comply. These laws range in subject from health and safety, labour rights, equality and discrimination, criminality to competitive practice. If a law impacts upon a business then the business has to comply. While some laws are not always popular and while in a democratic environment there are opportunities to challenge the system, the rule of law has to be observed. Companies that are regularly prosecuted for violations are not good corporate citizens.

An area that tends to create ethical dilemma for companies is competitive practice. Microsoft has emerged as the world's most successful software house but has faced numerous charges of anti-competitive practices in both the United States and Europe. The original case was brought about through unfair practices in packaging the company's dominant operating system with its web-browser. The argument presented was that packaging the two products together was to use the monopolistic position of the lead product, the operating system. Although the courts ruled in favour of the various governments, there are still arguments from proponents of the free market that all Microsoft did was to follow simple good business practice.

In the United Kingdom, the Enterprise Act, 2002, made it a criminal offence for individuals who dishonestly engage in one or more of the prohibited cartel activities: price-fixing; limitation of supply or production; market-sharing and bid-rigging. The cartel offence operates alongside the existing Competition Act regime, which provides for the imposition of financial penalties on companies that breach the Chapter I prohibition on anti-competitive agreements. The Office of Fair Trade (OFT) has recently searched the premises of twenty-two companies in Nottinghamshire, Leicestershire, Derbyshire and South Yorkshire as part of an investigation into allegations of collusive tendering for public and private contracts in the construction industry between 2000 and 2005. The investigation is currently being conducted under the OFT's civil powers under the Competition Act, 1998. The investigation may uncover behaviours in breach of the criminal cartel offence. If proven, these are a breach of the law, suggesting that individuals knowingly colluded in an unethical activity at the corporate level.

The law is varied and has many components. Companies often try to circumvent the law and even use the local law as an excuse to move their operations to environments where the laws are regarded as less onerous. While laws may vary from country to country, there is an increasing question as to what is acceptable behaviour from a company. A key question is how companies should operate when they have the option of working in environments that have lower legal expectations than in their home environments. Is the lowest acceptable standard adequate or should companies work towards maintaining the highest possible standards, as they have been forced to do at home?

Bribery and corruption

There is a propensity for companies involved in engineering to be involved in corrupt practices. The defence and arms sector along with construction have constantly appeared at the top of Transparency International Corruption Indices.

The temptation for companies to engage in corrupt practices is usually driven by opportunity to make greater profits. The ethics of those who support such behaviour has to be questioned. The consequences of corrupt practices are not restricted to the individuals perpetrating the actions but also to other stakeholders.

In the construction industry, the potential for corruption is accentuated by the size and scope of the sector, the value of which is estimated globally at some $3200 billion per year. Corrupt practices and lack of transparency in the contracting processes for the design and construction of large-scale infrastructure projects can have devastating consequences for economic and social development of the areas involved. Transparency International sees corruption in the construction sector not only plundering economies but also actually shaping them. Corrupt officials working with similar-minded contractors can steer social and economic development towards large capital-intensive infrastructure projects that can be exploited. Corruption can also lead public spending towards projects that are environmentally destructive.

The Lesotho Highlands Water Project was a dam construction venture, which was partly funded by the World Bank, that became notorious for the corruption scandal that subsequently emerged. The project forms part of a water control system comprising at least five dams and pipelines, controlling rainfall in the Lesotho mountains and supplying fresh water to South Africa. The value of the project is estimated at around $8 billion. The Lesotho government initiated an enquiry that ultimately led to the chief executive of the Lesotho Highlands Development Authority being convicted of taking bribes from some members of the consortium constructing the project. The case also implicated members of the international consortium in providing the bribes. The World Bank indicated that it would in future ban any company found guilty of corruption from involvement in its future programme of work.

The World Bank has also taken positive steps to reduce corruption on its projects. Its Department of Institutional Integrity investigates allegations of fraud, corruption, coercion and collusion related to World Bank Group-financed projects. In addition, it investigates allegations of serious staff misconduct. It has investigated over 2000 cases since 1999. The investigations by the World Bank have led to the disbarring of companies and consultants that have been found guilty of fraud. This in turn has led to more transparency in projects in developing countries where the threat of exclusion from World Bank-funded projects is a significant deterrent.

Transparency International, following on from the World Bank actions, has recommended a EU-wide debarment, or 'blacklisting', system for corrupt practices on EU contracts. Debarment disqualifies companies found to have engaged in corrupt behaviour from bidding on public contracts, thereby preventing corrupt firms from benefiting from EU contracts. This serves as a powerful deterrent to illegal behaviour. Active lobbying by other NGOs led to this idea being forced on to the EU agenda.

In order to improve their images some companies have decided to be more open about reporting corrupt practices. Shell commissioned a report into its practices in Nigeria (Shell 2002). The report indicated that the company had contributed to corruption in Nigeria. As part of its social reporting process, Shell also initiated an internal audit into bribery, facilitation payments and fraud that could be proved. In 2004 some 139 staff and in 2005 some 107 staff were found guilty of such activities (Shell 2005). This was an important move towards transparency. This action is maintained as part of Shell's business integrity process, which highlights the training of staff in the use of intermediaries and facilitation payments to contractors as part of its everyday business activity. This suggests that either the company realizes that the sector in which it operates is prone to a culture of bribery and corruption or it is attempting to root out corrupt individuals from within its organization. Shell is recognized to have a high profile and over time its reputation has been seen to suffer as a result of its size and worldwide involvement in a complex industrial sector.

Questions

Imagine that you are on the board of a medium-sized engineering company. You have been asked to draw up a policy on bribery and corruption, including practical guidelines. What would you include in that policy?

Political issues

Should a company be involved in the political process? There is a widely held view that it is certainly in the best interests of companies to lobby politicians to support and protect their industries. All types of associations, both commercial and not for profit, form influence groups that attempt to protect their member's interests in the political process. The simple premise is that everyone should have access to the political process in the country in a democratic environment. The ability to inform the legislative process is acceptable as decisions should not be made in isolation and should reflect realities and opinions of society. Why then is the issue of corporate involvement in politics an ethical consideration?

A key principle in a democracy is to govern for the many and not the few. Companies and their associations are often seen as having disproportionate influence because of their lobbying activities. Lobbying is the practice of promoting a point of view with a goal of influencing that point of view. The concept of lobbying is open to all and both commercial and public advocacy groups engage in the activity. Commercial associations through their members have the ability to generate large amounts of money for lobbying activities and accordingly face accusations of having undue influence. Lobbying has become an industry in its own right with over 35 000 registered lobbyists in the US Congress and 10 000 employed in the EU. This industry has to deliver results to those that finance and support them, and their member's expectation of influence on the political process is paramount. As agents of commercial organizations and associations, lobbyists can operate at a very simple level by providing research and briefing notes to politicians. In countries where there are limited restrictions on funding for political activity, companies can make contributions to political parties. In the United States, lobbying activity can be taken up directly in support of a political cause or activity, normally through a political action committee (PAC). An example of this would be access to mainstream media where a wealthy PAC can fund adverts for their chosen cause or politician. Recent scandals in the United States have led to a re-evaluation of the lobbying process and the introduction of the Legislative Transparency and Accountability Act of 2006. Whilst not yet altering the lobbying landscape, it recognizes the problem of lobbying and the companies that engage in the practice.

In the United States, lobbying is the most visible sign of influence whilst in the United Kingdom, there is a growing reliance by the main political parties to find sponsors to cover expenses. Many companies provide donations directly to political parties or through foundations. A concern is that a number of listed companies donate to political parties. The professional manager of a company should be acting in the interests of all its shareholders and questions are now asked about whether such donations are good governance. A stakeholder-based approach to governance would mean that a company would balance political influence with similar status given to other stakeholders.

In similar cases in countries across the world, companies have engaged in political activities through funding political parties or their affiliates. Often, high levels of political corruption have been financed by companies chasing lucrative contracts. In April 2002, BP changed its policy and stopped making corporate political donations anywhere in the world. There were no donations to UK or other EU political parties or organizations during 2005. In the United States, a similar approach was used but did not preclude individuals joining PACs. This did not infringe on individual rights but specifically excluded BP being directly aligned to such donations. The company monitors compliance with this policy via third-party certification process. The oil and petroleum industry has faced numerous calls to remove itself completely from the political process – particularly in developing countries. Exercising undue influence with political structure to gain concessions is a problem that exercises CSR thinking.

The more companies try to exercise political influence the more likely they are to be accused of trying to gain undue influence, and companies that are active in the political arena are scrutinized intensely not only over their political activity but in all their commercial enterprises.

Whistleblowing

A key part of improving business integrity is to create an environment that allows people the right to reveal unacceptable practices. Cynthia Cooper of Worldcom and Sherron Watkins of Enron, who exposed corporate financial scandals, are among the two most well-known whistleblowers. Jeffrey Wigand is also a well-known whistleblower for exposing that tobacco companies knew about the addictive properties of cigarettes and the carcinogenic ingredients in tobacco. Whistleblowing is not about vindictive actions by disgruntled employees but rather a means for companies to gain information and take corrective action.

Of course, whistleblowing is not easy for the individual concerned. There are core tensions in making such information available, either to the public or to superiors, which make it a real ethical dilemma, including the effects on:

- Peer professional relationships.
- Relationships with management. In these areas whistleblowers commonly find that they are victimized.
- Family relations. Whistleblowing can affect the family, either through subsequent stress or even through loss of employment.

A common argument against whistleblowing is that it is disloyal and that an engineer or manager has the duty of loyalty to the firm. Whistleblowing can also lead to negative effects on innocent colleagues and firms, not just on those involved in unethical or illegal practice. For Arthur Anderson, the Enron case meant the end of the whole organization.

The argument from the point of view of loyalty is difficult because it assumes an unquestioning loyalty demanded of the engineer/employee by the employer. However, loyalty cannot preclude the questioning of practice. As the Challenger case showed, there is an equally pressing obligation on the part of the engineer to maintain professional autonomy and to communicate the truth, which may have major safety implications for other stakeholders. It could be added that it is in the company's interest to learn of unacceptable practice, precisely so that it can avoid major consequences to the company and the employees.

The core question then becomes one of how the company enables whistleblowing so that it recognizes the needs of all the stakeholders and appropriately values the whistleblower. In England, the Public Interest Disclosures Act, 1998, provides a degree of legal security in that dismissal is forbidden for certain disclosures (Borrie and Dehn 2002, 105). Behaviour that falls outside these provisions still remains problematic especially where standards of ethical behaviour of firms have not been established as a code of practice. The individual whistleblower behaving in an ethical manner remains exposed in these situations.

Hence, it is important for the company to have in place both a culture that encourages well-thought-out whistleblowing and a process that enables this. Shell has made the provision for whistleblowing available as part of its campaign to improve its business integrity.

Shell Helpine (www.compliance-helpine.com/Shell.jsp)

The coherence of Shell's whistleblowing policy involves:

- Establishing the importance of reporting to the reputation of the company and its brand.
- Making the helpline available to anyone inside or outside the organization.
- Providing a set of criteria with which to measure unacceptable behaviour. This involves the Shell General Business Principles, which included good practice in competition; business integrity; health, safety, security and the environment; relating to local communities; compliance.
- Ensuring that the helpline is confidential and professionally managed. This requires that any complaint is given full attention and respect, whilst also being properly assessed.

Such a system needs to be well managed and may be open to abuse, unless there is some transparency in it to balance confidentiality. Hence, Armstrong et al. (1999) argue that clear records should be taken and that the reporter should have the right to appeal to an external authority if he or she believes that the issues raised have not been met. It would be in the interest of the group to have independent advisers who could remain part of the circle of confidentiality. Shell have such a group involved in monitoring their reports.

Whistleblowing is a particularly good example of the way in which ethics has moved away from focusing on the individual and on individual responsibility for working out dilemmas. Instead, it sees the responsibility of reporting bad practices as shared by the company and all employees or consultants. On the basis of the ethical framework mentioned in Chapter 3 this ensures that data are effectively gathered. Values then become central to the process as the issue is compared with the stated values and principles of the organization. The responsibility for taking the situation forward is then shared by the initial whistleblower and the management. Careful attention is paid to how the problem is corrected, such that the whistleblower is protected. It is then in the interest of all concerned to get such a process right. Having a confidential external element also provides a way of handling issues that might not be covered by the stated values.

This approach to ethics again has much to do with the culture of the company, and how it develops in response to different challenges. Armstrong et al. (1999, 56) suggest a useful checklist to back up such a process and one that can focus on maintaining good relationships. This checklist includes the following:

- Confirming that the risk to the public or fellow workers warrants action.
- Examining the motives for whistleblowing. Here a trusted colleague might be of help in reflection. There is a danger of heroic motive, with the image of the whistleblower as defeating the 'evil' organization, or a revenge motive.
- Evidence should be checked, verified and recorded.
- Stating clearly objections to the practice involved, rather than the people. Where objections become person centred there is the danger of losing objectivity.
- All company procedures should be followed by the potential whistleblower, and only if there is evidence that the company is not responding should external disclosure be considered.
- Carefully attending to the outcome of the process such that the whistleblowers are treated justly.

Supply chain management

The nature of modern business has moved away from the all-encompassing organization to a more specialist company. Many companies are simply organizers of supply chains and alliances to deliver their products. As economies of scale, cost structures and specialist skills are more widely dispersed, many companies have international partners within their supply chains. Modern methods in business now make the members and partners within the supply chain effectively part of the firm. A supplier, subcontractor, specialist or consultant that contributes to the organization is part of it. In most cases, outsiders are used because they create value for the company. Besides creating value, they also have an influence on the reputation and the overall corporate ethos of the firm.

The most visible cases that demonstrate the impact of supply chains are in the clothing and sporting goods industries. Many well-known brands used 'offshore contractors' in low-cost economies to produce their products. Investigations indicated that in many cases the workers in these offshore companies were working in sweatshops in poor conditions with extremely low wage levels. In some cases the use of child labour was also identified. Well-known sport goods manufacturers who were trying to develop good corporate profiles in their home markets were suddenly exposed as uncaring and not exercising care or good judgement with the choice of their subcontractors. The defence that local practices and conditions were the subcontractor's responsibility was no longer acceptable. The subsequent outcry and loss of sales changed minds. The lesson is that the supply chain is inextricably linked to the main organization.

It is common practice for suppliers and specialists to have an assessment prior to being awarded a contract. These assessments normally relate to skills, capability, financial strength, equipment, staffing, quality systems and so on. There are generally no explicit references to ethics or social responsibility issues. In the construction sector it is commonplace for there to be extensive pre-qualification in the area of health and safety. As the regulatory environment for health and safety has tightened up this has become a more important issue. Most lead contractors have joint liability for the safety of workers who belong to their subcontractors. The globalization of production means that companies have to be more aware of the labour practices, environmental concerns, working conditions and so on of their partners. A partner behaving in an unethical manner undermines corporate goodwill. Good CSR practice indicates that the behaviour of the members of a supply chain or alliance is important. The values that a company develops for its corporate identity have to be reflected in its supply chain. The principles of corporate responsibility established within this chapter should be reflected in the supply chains.

One of the sectors that has faced the maximum criticism for its supply chain has been clothing manufacturing. Sweatshop conditions have been identified with a number of leading brands. An initiative that has been proposed is ethical labelling. This process would mean indicating on the label of every garment the factory and address of its origin. This would indicate a degree of openness about supply chains as factories could be easily identified and poor practice exposed. An extension of this idea would be for engineering contractors and manufacturers to identify where their products were sourced. A scheme for the identification of the use of illegally logged timber from rainforests is in operation. This type of labelling activity forces more companies to consider their sourcing and supply chains.

Community involvement

The development of a community strategy is concerned with the orientation of the company to key stakeholder influences. Corporate involvement in the community is not only geography and project specific but also requires a long-term outlook.

The community strategy is aimed at issues that influence the firm irrespective of the project or location. The community policy aims to support activities that are of strategic interest to the firm, as well as its norms and values. Other influential factors are general social, economic, political and environmental issues. This part of the emerging strategy is concerned with the long-term approach needed to make community policy meaningful. It is easily implemented because it relates more directly to the corporate ethos of the firm. The values a firm adopts will provide the springboard to community-related activity. A company committed to education and training of its workforce would be able to translate these activities into a community policy. A literacy initiative, skills development programme and so on may follow as a community strategy. A great deal of this activity will emerge from the broader CSR strategy.

The starting point for the community policy is vested in the corporate goals and values of the company. These are specifically the value system of the organization and its CSR goals. The strategic objectives of the company should include a clear altruistic commitment. All subsequent action can then be evaluated against these aims. Without a value-based system the actions that a company takes will be seen as nothing more than public relations. The policy should not only be community based but also be part of a broader corporate responsibility ethos. This removes the charges of image building rather than having concerted responsibility values.

Core to this is the identity of the company, seen not as the centre of operations around which related interests operate, but rather as itself a stakeholder in society, which shares community concerns – a corporate citizen. Where this also involves employees working in voluntary capacities in the community there is also added value to staff development.

An essential part of the community strategy is the commitment of resources to the policy. This may take the form of finance, staff or equipment. The commitment of resources for what can be seen as intangible returns is often the most difficult decision to make. Inadequate resources and the lack of genuine commitment to the policy will certainly lead to failure of any scheme. If the policy is linked to the corporate goals then the schemes are more likely to have adequate resources allocated to them. Involvement in costly schemes that do not satisfy community needs is futile and often counterproductive to the aims of the policy ideals. Consultation with interest groups on the actions and activities to be undertaken is important. The practicalities of the implementation, the problems and the possible outcomes are more likely to be identified during this process. The use of third-party organizations, such as Business in the Community, United Way or other charities, can help shape and frame community policy. The approach undertaken by the organization must fit into its overall values. A good example of working with such groups is the Wates Foundation.

Case 6.1
The Wates Foundation

The Wates Foundation was founded by the Wates family who are associated with the Wates Construction Group. The mission of the foundation is to believe that the quality of life in society can be improved by informed and independent intervention to alleviate distress, deprivation and social exclusion. The foundation aims to alleviate distress and improve the quality of life by promoting a broad range of social priorities with a core area of special focus. There is emphasis on the physical, mental and spiritual welfare of the young and disadvantaged aged 8–25. Grants are concentrated on projects in the Greater London area with a preference for South London. The foundation focuses on activities that are related to social and community improvement. The activities of the foundation do not relate to the activities of Wates Construction but are part of corporate responsibility activities of the commercial organization. The Wates Foundation works towards helping and developing specific communities. Its actions are designed to help improve a particular location. In this case the activities of the charitable foundation are not related to the construction activity. The Wates Construction company has its own corporate responsibility programme but recognizes the role played by the charitable trust in helping communities. The Wates approach does show a recognition of the broader problems faced by communities and the need to tackle them. The philosophy is that a better society will improve everyone.

The traditional form of community interaction has always been charitable giving or work in a local community. The UK house builder, Persimmon, has had its corporate headquarters in the city of York for thirty years. It has made a charitable pledge to the York Minster development campaign for ten years. York Minster is the symbol of the historic city and offers Persimmon the opportunity to be involved with a project that is at the heart of civic pride. It also extends its charitable activity towards training schemes for apprentices for the Minster. New apprentices are required for the traditional skills that are needed for the repair and conservation of the Minster. Skills development will help the upkeep of the Minster in the long term. However, this does not exempt the company from facing rigorous scrutiny when it operates within the city. It still faces the criticisms and planning objections that are the normal part of a critical review in any community. It was also involved in a controversial scheme to redevelop the stadium of the local football team as a commercial housing complex. This type of situation presents the kind of tension that any company may have to handle: seen to be doing good for one part of the community, yet perceived to be damaging another part of the community.

Labour and human rights

The International Labour Organization Declaration on Fundamental Principles and Rights at Work (1998) is an expression of the commitment by governments, employers' and workers' organizations to uphold basic human values – values that are vital to our social and economic lives.

The declaration covers four areas:

1. Freedom of association and the right to collective bargaining
2. The elimination of forced and compulsory labour
3. The abolition of child labour
4. The elimination of discrimination in the workplace

The declaration enshrines the right of workers and employers to freely form and join groups for the promotion and defence of their occupational interests. This right goes together with the right to freedom of expression. It is the basis of democratic representation and governance. Those concerned need to be able to exercise their right to influence matters and their voice needs to be heard and taken into account on matters that directly concern them. The right to form associations to protect both parties' interests are protected by the declaration. It also sets out the right to collective bargaining, striking, collective decision-making, dispute resolution, mediation and arbitration in worker–employer relations. However, for the system to work there must be balanced laws that allow freedom of association and collective bargaining to thrive.

Circumstances often force people to give up their freedom as a result of economic, cultural or social circumstances. It is from these scenarios that exploitation of labour can occur. Forced labour according to the ILO is something quite distinct and it occurs where work or service is exacted by the State or individuals who have the will and power to threaten workers with severe deprivation. Actions such as withholding food or land or wages, physical violence or sexual abuse, restricting peoples' movements or locking them up are deemed to be the actions that constitute forced labour actions. People trafficking can give rise to forced labour as people are often bonded to criminal operations. Forced labour is likely to operate in the underworld in developed economies but companies using extended supply chains across the globe leave themselves open to unknowingly supporting promoting forced labour if they do not scrupulously vet their sources. A fundamental concern for all organizations' CSR activities would centre on this area of protection of human rights.

While the declaration does not preclude child labour it does believe that the world should be working towards the abolition of child labour. While most engineering companies will condone the use of child labour, it is possible to source products that have had a child labour input. The incidence of child labour is particularly prevalent in Africa and in the mining sector. This does not occur in the formal sector with reputable mining houses but in small-scale labour-intensive operations in gold and diamond mining. The sale of diamonds in unregulated markets in particular is a lucrative source of income to groups pursuing armed conflict in Africa. Beyond Africa, companies need to make sure that their supply chains are free of child labour. Child labour is examined more closely in Chapter 8.

Discrimination at work can occur in various forms. It can affect men or women on the basis of their sex, or because their race or skin colour, national extraction or social origin, religion or political opinions differ from those of others. Discrimination can also target the disabled, or those with illnesses such as HIV. Discrimination at work denies opportunities for individuals on an equal basis. Many countries have laws that protect people from discrimination but it is the attitude and culture of individual companies that make the difference with regard to discrimination in the workplace. In the United States, the constitution protects individuals from discrimination. The culture of the organization is fundamental to the prevention of discrimination in the workplace. Organizational culture is created by the values and attitudes of its senior managers.

The effect of any firm's actions, directly or indirectly, on human rights can have major implications for investors. A good example of this is the Norwegian Government's Oil Fund.

Case 6.2

The Norwegian government's Oil Fund ($240 billion), the world's largest pension fund, has indicated that it will no longer invest in the retailer Wal-Mart. On a recommendation from the fund's ethical council it was decided to exclude Wal-Mart for the serious and systematic violations of human and labour rights. The ethical council found that Wal-Mart had broken international norms including employing minors, allowing hazardous working conditions at suppliers, blocking unionization and discriminating against women in pay terms. It looked at its operations in the United States and Canada and suppliers in Nicaragua, China and El Salvador. The Norwegian finance minister indicated that 'these companies were excluded because in view of their practices, investing in them entails an unacceptable risk that the fund may be complicit in serious, systemic and gross violation of norms' (Guardian 2006). The ethical position taken by the Oil Fund indicates that investors are becoming more sophisticated in their approach to evaluating companies.

Another example of a more contentious nature is that of Caterpillar. Caterpillar is one of the world's leading earthmoving equipment manufacturers. Like all progressive organizations, it has a worldwide code of conduct for its operations. In early 2006 Caterpillar faced a dilemma when it was suggested that its equipment had been used in the destruction of homes in Palestine. It has faced lawsuits and other protest actions in the manner in which its equipment has been used. The Church of England at a General Synod in 2006 chose to withdraw investments in the company. This raises major questions about how far Caterpillar could have been said to have infringed human rights by extension.

Questions

Is a company ethically responsible for the actions taken by its customers? Should it continue to sell its products to the said client?

With some colleagues imagine that you are the board of Caterpillar. How would you respond to the decision of the Church of England? If you continue to sell your products to Palestine, how would you justify this?

How might this case differ from Anglo American's or Interface's view of CSR (see Chapters 7 and 8) where they argue the importance of checking that subsidiaries share something of their values? Can companies demand that customers share their values? If so, how could this be justified?

Human rights form the basis of many of these issues and they will be considered in more detail in Chapter 8.

Environmental concerns

The environment is very firmly on the ethical map. The Montreal, Rio, Kyoto and Johannesburg summits have placed environmental concerns near the top of the agenda for international concerns. Organizations are now being assessed in more detail over their environmental performance. More companies are producing environmental reports and monitoring their environmental performance. The cost of poor environmental performance in terms of fines and loss of reputation is increasing. The Norwegian Oil Fund, mentioned earlier, has also indicated that it will no longer invest in the US mining group Freeport-McMoran Copper and Gold for environmental reasons. This decision influenced the share price of the company.

The environment is an important part of most company's CSR activities. The environmental performance of the company is increasingly scrutinized by society.

The growth of organizations such as Greenpeace indicates that public awareness of environmental issues is becoming more relevant. The specific issues related to environmental ethics are dealt with in Chapter 7.

Health and safety

The European Agency for Health and Safety in its tenth anniversary brochure (2006) highlights the health and safety concerns for some190 million citizens:

> This isn't just in the interests of the individuals who work in Europe. It is a prerequisite for unlocking the EU's social and economic potential, as identified in the Lisbon agenda. You only have to look at the human costs of some of the most common occupational safety and health problems in Europe to appreciate the importance of this issue. Every two hours, for example, someone is killed in an accident at work in the EU. Moreover, Europe loses an estimated 550 million workdays per year through occupational ill health and accidents. Contrary to popular opinion, these risks – and the associated social and economic costs – are not confined to traditional 'heavy' industries. Virtually every sector, from finance and agriculture to healthcare and the entertainment industry, is affected. Musculoskeletal disorders, noise-induced hearing loss and work-related stress are just some of the consequences.

Health and safety is a fundamental part of CSR and is concerned with the health, safety and protection of employees and employers as well as suppliers and customers that interact with the business. ILO protocols usually mean that most countries do have basic statutory provisions for health and safety. The enforcement of regulatory provisions varies across the globe and engineering companies tend to apply these regulations in a manner consistent with the local level of enforcement.

In July 1988, the Piper Alpha oil platform, overloaded and ill-maintained, exploded, killing 167 people. The Cullen Inquiry into the disaster found Piper Alpha's operators, Occidental, guilty of seriously inadequate safety procedures. The profound impact was that health and safety procedures were dramatically improved in one of the most dangerous engineering sectors. A key change in the regulatory environment was the Offshore Installations (Safety Case) Regulations, 1992 (amended 2005). They require the operator/owner of every fixed and mobile installation operating in UK waters to submit to the HSE, for their acceptance, a safety case. Safety cases are also required by the HSE at the design stage for fixed installations and when installations are involved in temporary drilling or well activities or are decommissioned. The safety case must give full details of the arrangements for managing health and safety and controlling major accident hazards on the installation. It must demonstrate, for example, that the company has safety management systems in place, has identified risks and reduced them to as low as reasonably practicable, has introduced management controls, provided a temporary safe refuge on the installation and has made provisions for safe evacuation and rescue. While there are still fatalities in the offshore industries, there is now a serious commitment to improve the safety record. Corporate social responsibility behaviour in this sector is heavily dependent on the need to deliver a safe work environment. Once again it took a major incident to begin to change the culture of an industry.

Health and safety law breaches are the most widely visible area of breaches in the law. Between 1999 and 2005 the lowest single-year work-related fatality figure was 249. There is a view that one work-related death is one too many. In pursuit of duty holders of companies that might be responsible for breaches in Health and Safety, less than 50 per cent were prosecuted. In 2004/2005 there were 712 duty holders prosecuted with 673 convictions with total penalties of around £12 600 000 at an average of £18 700 (HSE 2005). A business environment that is safe as possible for employees is a key component within CSR philosophy.

Are health and safety laws responsible for poor practice in the engineering sector? In the United Kingdom it is very difficult for a company to be tried for corporate manslaughter and gain a prosecution. As Thompson (2006) notes,

> The prosecution has to prove beyond reasonable doubt not only that an unlawful killing took place but that firstly, a responsible person within the company controlled the activities, resulting in the death of the individual (the controlling mind); and secondly that person was grossly negligent so as to be responsible for the death and could be prosecuted, as an individual, for the offence of manslaughter.

The company is only guilty of corporate manslaughter if there is gross negligence and a directing mind has been identified as the embodiment of the company. The principle of aggregation, which attempts to identify collective responsibility, has not been applied in UK corporate manslaughter cases. In the USA, the principle of aggregation applies as the attitude of the US courts is to identify guilt of the company rather than the individuals. In the United Kingdom, corporate manslaughter convictions are primarily related to small business where the owner/manager was easily identifiable. In the Lyme Bay Tragedy (R *v* Kite and OLL Ltd (1993)), Kite was identified as the controlling mind and became one of the few directors of a company to be given a custodial sentence. For larger companies the conviction rates are very small and many companies seem willing to plead guilty for breaches in health and safety. Fines do not ultimately deter large companies from poor performance. The law tends to fine companies but rarely makes individuals responsible.

The right to work in a environment that is safe is a fundamental right of the worker. Responsible employers try to reduce the level of danger to their employees. An ethical company will try to reduce the danger to its employees where it is reasonably practical.

Consumers

The products that a company manufactures represent the company. The manner in which a product is produced, sold, marketed and supported is important. A significant issue is the safety of the product in relation to its use. A product that is unsafe and is knowingly sold is unethical. In a similar vein, if the company discovers that the product is unsafe and does not inform the public then this is also unethical.

In 2001, Ford recalled 50 000 of the 2002 Explorer SUVs. This vehicle was meant to be safer than an earlier version that had a poor safety record. However, flaws meant that the vehicle was recalled. A year earlier Ford and its then tyre supplier, Bridgestone, were involved in product recall with the Explorer. Both companies attempted to blame each other and a protracted legal dispute ensued. The car industry has a record of a number of similar cases including those of the Pinto and the Corvair.

In all the cases mentioned there was loss of life before the companies investigated the defects. Ultimately, if the product is defective then there is a need for the company to accept responsibility and let the consumers know.

The German car manufacturer, Mercedes Benz, withdrew from sale its new A-Class small car in 1997 after fears about its safety and stability. In one road test, a production model toppled over while making a sharp turn to avoid an obstacle. The A-Class model underwent three months of redesign, and it is estimated that the process cost Mercedes Benz at least $150 million dollars in lost orders. Mercedes had been accused of rushing a new car on to the market without thorough tests. However, the action taken by Mercedes before the car became publicly available helped it retain its reputation for engineering excellence. By acting promptly the situation was rescued.

Every engineering company has to be aware of the goods or services it produces. They have to be manufactured to the standard that is acceptable. The producer of any engineering product, or provider of any service, owes a duty of care to the users of its products and those that come in contact with the products. The principle of the duty of care to consumers of products was founded in the Donahue *v* Stevenson case (House of Lords 1932). In this case, the consumer did not have a contract with the producer but it was held that the user was still owed a forseeable duty of care. The concept of duty of care to the consumer is fundamental to ethical considerations.

Conclusions

A company is not a person but it can and should be a corporate citizen. The manifestation of its ethical behaviour is its CSR. Its CSR policy and implementation set the ethical tone of the organization. However, the CSR activity cannot be separated from the actions of the individuals within the firm.

It is not possible to have excellent CSR programmes when the behaviour of the leaders of the organization is not congruent with principles, policy and practice.

This chapter has suggested:

- The breadth of CSR. This continues to develop. We noted that this can contribute towards the social ethics of society. This is increasingly clear in former Eastern Bloc countries such as Croatia. Gregory and Tarfa (2004) note the way in which CSR, developed through good public relations networks, has 'a critical role in promoting democratic accountability and in building communities'. In a nation that is in transition from communism to the free market, this directly contributes to the development of shared values. In a nation that is still recovering from war it contributes to social cohesion. Links to colleges and other civic institutions further enhance this process. Gregory and Tarfa note that the importance of CSR in this context is not the stress on developing values and ideas but rather the action orientation. This embodies and articulates values more effectively and builds up the trust required to develop social capital. In turn this can contribute towards national reconciliation.
- The different motivations for developing CSR. Like professional ethics it is legitimate to see the development of CSR as a matter of self-interest for the company. Self-interest does not preclude ethics. On the contrary, the company needs to balance the interests of and responsibilities to itself with the interests of and responsibilities to wider society. The company also has an interest in the social and physical environment in which it operates.
- CSR as a function of creative partnership. Partnership enables dialogue on values, through reflection on creative action. It also enables the appropriate and most effective response to the interests of the different stakeholders.

Recent developments have taken this further with the Globally Responsible Leadership Initiative (2005). Arising from the UN Global Compact this argues for:

- Improved leadership training that takes seriously the ethical role of leaders.
- A holistic approach to ethics and CSR that enables the development of a responsible culture in business.
- The focusing of CSR and ethics in a global context.

It is to the global and environmental aspects of this that we now turn.

Questions

The developing view of CSR is broad. What are the limits to CSR? Where would you draw the line over CSR involvement, and what criteria would you use to decide?

Imagine that you are in charge of a small business. Would it be possible to develop an effective CSR policy that takes into account the broader view of CSR?

References

Books

Andriof, J. and McIntosh, M. (2001). *Perspectives on Corporate Citizenship*. London: Greenleaf.

Armstrong, J., Dixon, R. and Robinson, S. (1999). *The Decision Makers: Ethics for Engineers*. London: Thomas Telford.

Borrie, G. and Dehn, G. (2002). Whistleblowing: the new perspective. In *Case Histories in Business Ethics* (C. Megone and S. Robinson, eds) pp. 96–105, London: Routledge.

Cadbury, A. (1992). *Report of the Committee on Corporate Governance*. London: HMSO.

Daugherty, E.L. (2001). Public relations and social responsibility. In *Handbook of Public Relations* (R.L. Heath, ed.) pp. 389–401, Thousand Oaks, California: Sage.

European Commission White paper (2002). *Promoting a European Framework for Corporate Social Responsibility*. Luxembourg: Office of Official Publications.

Gregory, A. and Tafra, M. (2004). *Corporate Social Responsibility: New Context, New Approaches, New Applications: A Comparative Study of CSR in a Croatian and a UK Company*. Paper given at International Public Relations Research Symposium, Bled 2004. www.bledcom/com/uploads/documents/Gregory.Tafra.pdf

Guardian (2006). *Biggest Pension Fund Boycotts Walmart*. 7 June, p. 22.

Higgs, D. (2003). *Review of the Role and Effectiveness of Non-executive Directors*. London: DTI.

Solomon, J. and Solomon, A. (2004). *Corporate Governance and Accountability*. Wiley. London

Sommer, F. and Bootland J. (2004). *Engage. How to Deliver Socially Responsible Construction*. A client's guide, Ciria C627. Ciria London.

Thompson, P. (2006). Corporate killing and management accountability. *New Law Journal*, 156 NLJ 94.

Turnbull, N. (2005). *Internal Control: Guidance for Directors on the Combined Code (Rev.)*. London: Financial Reporting Council.

Reports

British Petroleum Sustainability Report (2005). London: BP.

Globally Responsible Leadership Initiative (2005). UN Global Compact.

Health and Safety Offences and Penalties 2004/2005 (2005). A Report by the Health and Safety Executive. London: HSE.

ILO (1998). *Declaration on Fundamental Principles and Rights at Work*. 86th Session, Geneva: ILO.

Shell Nigeria Report (2002). London: Shell.

Shell Sustainability Report (2005). Social data 1997–2005. London: Shell.

7 *Environmental ethics and sustainability*

Introduction

The earth's natural resources, air, water and land are becoming increasingly polluted and abused at alarming rates because of human activities, resulting in an estimated 100 species becoming extinct every day (Wilson 1989). This rate of extinction could be accelerated by rapid climate change caused by increasing carbon dioxide levels in the atmosphere from industrial-scale burning of fossil fuels. The expansion of human settlements and intensification of agriculture have led to the destruction of natural habitats and ecosystems, such as tropical and temperate rainforests, freshwater lakes and streams and coral reefs, on a global scale. This situation raises questions about the social and environmental responsibility of business and the engineer.

The rise in green thinking and sustainability concerns since the early 1970s has been accompanied by the development of environmental ethics. Environmental ethics concerns itself with the ethical relationship between human beings and the natural environment in an attempt for both to co-exist without compromising the future of either by assuming that moral norms govern human behaviour towards the natural world (Des Jardins 2006).

In this chapter we will examine:

- Environmental ethics
- Sustainability
- How business can respond to these issues, including the development of audits
- The Chernobyl and Interface case studies

Environmental ethics

There are three primary schools of thought with respect to environmental issues, instrumental, axiological and anthropological, each placing different priorities on human activities and the natural world (Carson 1962).

Instrumental

The instrumental approach is a very human-centred approach in the sense that it views an improvement in humankind's relationship with nature as having importance and for humankind alone. Therefore the protection of nature only has value for humankind. This position has the consequence that if humankind has no instrumental use for nature then nature has no ground for protection.[1]

Axiological

In contrast, the axiological approach argues that nature has intrinsic value and that we should protect it because of this value. Hence, this approach seeks to establish what this intrinsic value consists of and where it comes from. The axiological approach to environmental ethics represents an alternative to instrumental environmental ethics (Chew 1999). In the axiological approach, the earth and its natural resources are perceived as having their own intrinsic value independent of human beings, in contrast to instrumental ethics, which regards the earth's natural resources as a means to achieve human ends. This approach is hard to sustain because it considerably diminishes humankind's important view of itself. It also allows for the unthinkable possibility of the earth and its flora and fauna existing quite comfortably without the requirement of a human presence.[2]

Anthropological

The anthropological approach is primarily concerned with what being human is all about, and it links the understanding of what the relationship between the human self and nature ought to be. This approach argues that humankind will engage in a relationship of respect with nature and how humankind can gain such a sense of the intrinsic value of nature by socially enhancing human life whilst ensuring that the environment is sustained, by keeping humans and nature in equilibrium. Many traditional Western ethical perceptions are human centred or anthropocentric in that they assign intrinsic value to human beings alone or they assign a significantly greater amount of intrinsic value to human beings than to any non-humans (Carson 1962). When environmental ethics emerged as a new subdiscipline of philosophy and ethics in the early 1970s, it did so by posing a challenge to traditional anthropocentrism. Firstly, it questioned the assumed moral superiority of human beings to members of other species on earth, and, secondly, it investigated the possibility of rational arguments for assigning intrinsic value to the natural environment in its non-human elements.[3]

---

[1] http://web.singnet.com.sg/~chlim/Chapter1.html (accessed on 20 February 2006).
[2] http://www.ccamain.com/pdf/earth_ethics.pdf (accessed on 20 February 2006).
[3] http://plato.stanford.edu/entries/ethics-environmental/#1 (accessed on 22 February 2006).

Neither the instrumental nor the axiological approaches are generally used as the foundations for environmental ethics. The instrumental approach lacks consideration for the environment and the axiological assumes that nature can be sustainable without the requirement of a human presence. The approach most widely used in environmental ethics today is the anthropological approach where equilibrium of respect for nature and human activities is searched for.

It was the work of Rachel Carson that carried modern-day environmental ethics forward and helped shape the way we presently look at environmental ethics. Carson has been called the founder of the environmental movement, primarily due to the publication of *Silent Spring* in 1962, which focuses attention on the problem of pesticide and other chemical pollution, and led to such landmark legislation as the US Clean Water Act and the banning of DDT in many countries throughout the world. Carson developed concepts for the long- and the short-term consequences to humans and the natural world of human activities, and the political and ethical implications of these activities. Concerns were raised about finding technological solutions to environmental problems. Science does have tremendous potential for helping us to understand and solve environmental problems; however, it is not purely objective and value neutral, and hence dynamic interactions occur between science, technology and society and these must be objectively monitored (Carson 1962).

The Gaia hypothesis and deep ecology

Not surprisingly, underlying some environmental ethics are belief systems and specific beliefs about the environment. One of the more prominent is the Gaia hypothesis devised by biologist James Lovelock (1979). He suggests that the earth should be seen as a single organism. Like any organism, it continues to attempt to maintain a healthy balance and fight disease. Some take this literally, seeing the human race as a part of this organism and therefore profoundly interconnected. In this light there is no sense in humanity trying to destroy or degrade the environment because it is the equivalent of destroying one's own body. The problem with this hypothesis is that if we view the human race as literally part of an organism then it takes away any real sense of moral autonomy, and thus responsibility.

Deep ecology focuses on the values of self-realization and biocentric equality (Vesilind and Gunn 1998). Neither of these is based in rational justification but is intuitively felt. Hence, deep ecology falls under the heading of existential awareness or even spirituality. Self-realization is an awareness of the self in relation to the wider universe. Given this awareness, humanity cannot think of itself as greater than any aspect of the universe, hence the belief in biocentric equality – all parts of the universe are equally valuable. The stress in deep ecology is on proper respect for nature, hence the importance in this view of maintaining the wilderness. Even the idea of stewardship is questionable in this light, because it implies intervention if not domination, rather than respect for what is.

Deep ecology requires a fundamental change in humanity's relationship with the environment. Ultimately, it will require massive depopulation, some arguing for as much as 90 per cent of humanity. This would require a complete change in terms of technology, and some critics suggest that it would necessitate a return to a hunter gatherer society. This in turn raises major questions of how it is to be achieved and who, other than the elite, might remain. Hence, whilst the underlying values of this view are important, the practical implications are less clear.

Questions

What is your view of the environment? How does that relate to your view of the engineering profession? How does it relate to you view of business?

Sustainability and ethics

Sustainability is a broad term. In some firms it has become a buzzword and a marketing tool through which projects are promoted. Where sustainability is part of this green marketing strategy, it is open to abuse. Hence, the term 'sustainability' needs to be critically defined, because products labelled 'green and environmentally friendly are increasingly saleable in today's markets'.

It is strongly believed in a number of ethical theories and religions that the purpose of the existence of man was to make him sustain the earth, helping him to tend and care for the environment. The principle of stewardship, for example, is central to the Judeo-Christian faith. It argues that the earth belongs to God and that humans fulfil the role of stewards (Robinson 1992). Therefore many people see the human race as being fundamentally interested in the well-being and sustainability of the earth (Oyem 2002). With respect to human activities in today's society, this can be achieved by taking a sustainable approach and by incorporating sustainable ethics in engineering practices. Sustainable ethics promotes morally accepted principles geared towards continued development of the environment for the present and future generations.

Sustainability ethics is an amalgamation of concepts of business and environmental ethics integrated with a focus on the CSR of an organization. Exponential population growth, depletion of natural resources, carbon dioxide emissions, energy waste and so on are not sustainable in environmental terms (Hurka 1992). Therefore sustainable practices should be embraced in a movement towards sustainable development and implemented in an ethical way. The transition to sustainable development and increased use of green practices within the construction industry requires an increased focus on the CSR of design consultancies and contractors in integrating the principles of ethics, social and corporate responsibilities into their daily actions as well as decisions on an individual and company level (Leao-Aguiar et al. 2006).

For companies to make the transition from conventional design and construction practices towards a more sustainable approach, there should be a greater emphasis on the social and environmental aspects of a project. This can be done by adopting a sustainable development programme that sets outs green goals and strategy for the company and the CSR and ethical values of top management within the organization to become more responsive about social and environmental issues (Leao-Aguiar et al. 2006). Although achieving these two criteria seem simple enough, one of the main challenges that businesses face when embracing the sustainable movement is the difficulties in striking a balance between economic activity and what is socially and environmentally sustainable.

The triple bottom line is the term used to capture the whole set of values, issues and processes that companies must address in order to minimize any harm resulting from their activities and to create increased economic, social and environmental values. Using a business strategy based upon the triple bottom line means that business goals and objectives should not be solely related to profits alone, but also with a concern to people and the planet. An organization's simultaneous pursuit of beneficial outcomes along three dimensions, economic, social and environmental, will aid in moving towards an ethically sound sustainable approach that can be controlled and monitored using triple bottom line accounting and reporting. The purpose in triple bottom line accounting and reporting is to create a higher level of organizational accountability and transparency to provide a more balanced approach for continuous improvement that acts as a learning tool for future business activity within companies. By measuring the triple bottom line performance of an organization, contributions can be made to a more sustainable community, which will contribute to a better understanding of sustainability. The technology used and quality of products desired by individuals and organizations are high, and achieving such high-quality comes at a cost. In real systems, there are limitations and the choice of the priorities and the allocation of resources become important; in most cases the high quality that is desired by clients appears to create lower levels in efficiency and environmental friendliness for projects. Hence, in the name of efficiency, sacrifices are made from quality. Therefore to resolve this issue ethically, equilibrium has to be found between sustainable quality and balanced efficiency, while incorporating social implications of projects.

Responsibility for future generations

The greatly increasing pressure on technology-based human activity has given rise to environmental insecurity and the depletion of natural resources. This raises the question about 'can we continue to operate civilization as we have been doing without spoiling or even ruining the future possibly for ourselves and our descendants?'

Sustainable development is an attempt to balance two moral demands that require serious ethical consideration. The first demand is for development, including economic development and growth. It arises mainly from the needs or desires of present generations, especially those groups that have a poor quality of life and urgently seek steps to improve it. The second demand is for sustainability, for ensuring that we do not sacrifice the future for the sake of gains in the present (Hurka 1992).

There are three main arguments against the present generations having responsibilities for the future generations. The first is centred on the ignorance of present generations with respect to conservation of the ecosystem for future generations. The second is based on the perspective that we have no obligation to bring future generations into existence because there are no particular people to whom responsibility is directed. The final argument focuses on the temporal location of future generations and concludes that the present generation cannot have responsibilities for people who will not exist for many years to come.

As humans, we usually accept responsibilities for our actions and their consequences, even if these consequences should occur at some point in the future. Therefore it is not the timing of the actions that humans should take as critical in determining human responsibility and accountability, but the course of detrimental action itself. Barring an unforeseen disaster, it is safe enough to assume that people will exist in the future and that they will be similar enough to us so that we can develop a good idea of their future requirements to sustain their existence. Taking this into account and knowing that the actions of the present generations can seriously influence the future generation's well-being, it is reasonable to assume that the future generations must be given some form of ethical consideration.

What do we owe future generations?

It has been established that we have a responsibility towards future generations to sustain the environment and the earth's ecosystem so that the existence of future generations is not compromised. The question arises as to what we owe future generations, and how do we decide where these responsibilities lie. The ethical basis of these responsibilities and their designations also needs to be established.

The obligation of present generations towards future generations by maximizing their happiness and reducing their suffering has led to some problems (Warren 1980). On utilitarian grounds it may be argued that we acknowledge responsibility for people of the future. However, those responsibilities can sometimes be overridden by the interests and needs of the present generations. Conversely, the interest of future generations might be jeopardized by the less important interests of the present. It is thought by some that the interests of the present population always override future interests because of the uncertainties in the needs and interests of future generations. Therefore the interests of future generations can be discounted on the grounds that uncertain and remote pleasure counts for less than certain immediate ones (Bentham 1897). The discounting of interests of future generations is done on the same principle as the reasoning that one unit of monetary value is worth more today than in some point in the future, thus the future value of money must be discounted in order to be equivalent to the present value (Arndt 1993). As a result, the practice of discounting future interests is common practice in economic analysis of environmental issues.

Another train of thought on the concept of what we owe future generations is the egalitarian view. This egalitarian view can take several forms, but all centre around the basic principle that considers each generation to have a duty to pass on to its successors a total range of resources and opportunities that is at least as good as its own. Those generations that enjoy favourable conditions of life must pass on similar circumstances of life to their successors: generations that are less fortunate have no such stringent obligation (Barry 1983). This egalitarian approach characterizes our duty concerning future generations not in terms of their well-being or quality of life but in terms of their range of opportunities. If present populations leave the opportunity for a high quality of life for successor generations, all is well and good. However, if the future generations misuse and squander the resources left for them then it is their fault, not that of the present generations, and does not mean that the present generations have failed in their obligatory duty. This approach does not require large sacrifices by the earlier generations. On the contrary, it never requires a generation to make sacrifices for the sake of generations that will be better off than it is.

The egalitarian view allows each generation to pass on an equal range of opportunities to its successors and a level of opportunities is achieved that is sustained through time. This view focuses on opportunities that fit the Bruntland Commission's view that 'development that is sustained is not quality of life as such, but the economic and other activity that permits quality of life' (Bruntland 1987). There are, however, two objections to the egalitarian view. The first objection is that it does not place excessive demands on earlier generations to make sacrifices for the sake of later generations because it assumes that earlier generations need do nothing for later generations. Earlier generations, however, have some duty to enable their successors to live better than themselves. The ideals of sustainability and maintaining a constant level of well-being through time may be attractive when starting from a high level of well-being, but it is not attractive starting from a low level of well-being. It is not sustainable for levels of misery to be consistently maintained. There may not be as stringent a duty to improve conditions for future generations as in the utilitarianism approach, but there is some level of duty.

The second objection is more abstract and concerns the way egalitarian views focus on comparative judgements. Imagine the scenario where one person is financially better off than his friend. Egalitarian views suggest that the reason why the better-off person should help his friend is simply because he or she is worse off. But how can this be a sufficient reason? Apart from any comparison with the better-off person, it is consideration for the worse-off person's condition that generates duty. In which case it is need, not equality, that provides the criterion for helping the other (Hurka 1992).

Reflecting on these objections towards the egalitarian view leads to a third view about the duty of the present with regard to future generations. Contrary to the utilitarian and egalitarian views, the duty of present generations is not to make the condition of future generations as good as possible or as good as our own, but only to make the condition of future generations reasonably good. This third view believes that each generation has a duty to pass on to its successor a range of opportunities that allows for a reasonable quality of life. If a generation can pass on a better range of opportunities, one that allows for a more than reasonable quality of life, it may be a nice or even admirable thing to do, but it is not their duty to guarantee this. This view rests on an idea for which economists have coined the term 'satisficing', which is a behaviour that attempts to achieve at least some minimum level of a particular variable but which does not strive to achieve its maximum possible value. Therefore this approach suggests that the present generations do not have to obsessively strive for the best possible outcome but should be content when they find one that is reasonably good, and is known as the 'satisficing' view.

The satisficing view lies midway between the egalitarian and utilitarian views with respect to earlier generations having a duty concerning later generations – that they should help enable their descendants to live reasonably good lives. However, if their own lives are not reasonably good, they may weigh a concern for their own interests against their duty to their descendants. This view helps regulate and moderate its demands on later generations. This is illustrated if you imagine that our present range of opportunities allows us a quality of life that is more than reasonably good. We do not violate a satisficing duty if we pass on a marginally smaller range of opportunities to our successors, as long as this range is reasonably large and The future generations can live above a certain threshold standard of living.

The defining concepts of the Bruntland Commission on the environment and development (1987) are meeting the needs of the present and enabling future generations to meet their own needs. Needs are not all that matter for a good life, but they come first before luxuries and have a certain priority. It is natural to define this priority in satisficing terms as people's needs are what must be satisfied if they are to have a reasonably good life, while wants and luxuries allow them an even better life.

Questions

How would you justify concern for future generations as an engineer?
How would you justify concern for future generations as a managing director?
Would there be any difference in the justifications?

Business ethics, sustainability ethics and corporate social responsibility

The World Business Council for Sustainable Development states that 'for the business enterprise, sustainable development means adopting business strategies and activities that meet the needs of the enterprise and its stakeholders today while protecting, sustaining and enhancing the human and natural resources that will be needed in the future' (Lehni 1992). Becoming sustainable makes business sense, and moves towards the '3 Ps' approach of focussing on people, planet and profits, instead of the traditional profit orientation. Even though some companies might have a good honest concern for the environment, their bottom line is to maximize profits, and this, ironically, is often the reason why companies are showing concern for people and the planet.

Greenwashing

In all this it is very easy for companies to get on the bandwagon of sustainability, precisely to offset any harm that might be done to their reputation and thus to the bottom line of profit. This has become known as 'greenwashing', defined by the *Concise Oxford English Dictionary* as 'disinformation disseminated by an organization so as to present an environmentally responsible public image'.

The 'Don't be Fooled Awards' (alternet.org/wiretap/15699) point to several major corporations that argue that they are doing just this. Exxon Mobil, for example, has contributed $100 million to the Global Climate and Energy (GCE) Project. However, at the same time, it has invested $100 billion in oil exploration. The impression of greenwashing was further reinforced when Exxon executives were disparaging about the GCE Project. Another oil company, Shell, is said to have patented a new environmentally friendly drilling fluid, which subsequently it did not use. A former Shell scientist argued that most of the 450 000 tons of toxic waste created by Shell in 2001 came from drilling fluids.

It is of course very easy to take aim at any major company, find problems and from that conclude that their environmental concern is a cynical front. Several things make this issue very hard to deal with:

- Many companies, such as oil and mining, are involved in businesses that could, by definition, be said to be environmentally harmful. Overuse of cars has contributed towards global warming, and oil companies enable and encourage this practice. Mining for minerals by its very nature scars the countryside.
- Such companies are one of several stakeholders, including governments, who have some responsibility for the effects of their industry.
- The size of such companies makes it very difficult to know the truth about any accusations of greenwashing. It may, for instance, be that bad practices occur in parts of the operation of which the central management is unaware.
- The dynamics surrounding attempts to develop sustainability in large companies can easily become polarized. It is easy to assume that MNCs' concern for profits excludes genuine attempts to develop sustainability, and thus for any progress not to be recognized, hence characterized as greenwashing. Behind this is the ethics of the perfection argument, i.e. that a company cannot be deemed ethical if it does not get everything right. Given the impossibility of getting everything right companies can become easy targets.

Any company then has to make clear that:

- It is not making a claim to perfection. As Ray Anderson (see the Interface case at the end of the chapter) notes, sustainability involves the development of awareness and the capacity to respond to need and is thus a continuing process.
- It is open to and has the capacity to respond to any external critique. This demands the development of genuine transparency.
- It will work in partnership with other initiatives. A good example of this is the Global Reporting Initiative (GRI). We will note this in more detail in the social reporting of Anglo American below.
- It must work carefully to develop corporate integrity. Integrity in this sense is the corporate version of the professional virtue. This is defined as developing a consistency between principles and practice across all contexts. It necessarily involves a process of continual learning.

A good example of an attempt to bring together principles and practice is the CERES principles. The Coalition of Environmentally Responsible Economies (CERES) initiated the GRI in 1997 and set out some basic principles in response to the Exxon Valdez disaster of 1989.

CERES Principles

1. Protection of the biosphere: provides for the elimination of pollution, protection of habitats and the ozone layer, and the minimisation of smog, acid rain and greenhouse gases.
2. Sustainable use of natural resources: commits signatories to conservation of nonrenewable resources, and the protection of the wilderness and biodiversity.
3. Reduction and disposal of waste: obliges signatories to minimise waste, to dispose of it responsibility and to recycle wherever possible.
4. Energy conservation: commits signatories to conserve energy and use it more efficiently.
5. Risk reduction: provides for minimising health and safety risks to employees and the public by using safe practices and being prepares for emergencies.
6. Safe products and services: seeks protection of consumers and the environment by making products safe and providing information about their effect on the environment.
7. Environmental restoration: accepts responsibility for repair of environmental damage and compensation for those affected.
8. Informing the public: obliges management to disclose to employees and the public, information about environmentally harmful incidents. It also protects employees who blow the whistle about environmental of health and safety hazards in their employment.
9. Management commitment: commits signatories to provide resources to implement and monitor the principles. It also means that the CEO and the company board will be kept abreast of all environmental aspects of the company's operations. The selection of directors will give consideration to commitment to the environment.
10. Audits and reports: commits signatories to an annual assessment of compliance with the principles that it will make public. (Grace and Cohen 2005, 154)

In the case of Chernobyl the contrast with such principles was very clear.

Case 7.1
Chernobyl

Reactor Four in the Nuclear Power Plant at Chernobyl was to be closed down for routine maintenance on 25 April 1986. Taking advantage of this, the staff were to run a test to see if there was enough electrical power to operate the crucial functions in the event of a crisis.

Halfway through the operation the electric load dispatcher prevented any further closedown. The emergency core cooling system was switched off and the reactor carried on at half power. Soon afterwards the supervisor agreed to a further reduction in power.

For the test the reactor was to be stabilized at 1000 MW prior to closedown. However, the power fell to about 30 MW, due to operational error. The operators attempted to increase the power by freeing all the control rods manually and in the early part of 26 April the reactor stabilized at 200 MW.

Soon afterwards the staff had to withdraw most of the rods due to an increase in coolant flow and a drop in steam pressure. This made the reactor unstable, with the staff having to adjust controls every few seconds to maintain a constant power.

The staff reduced the flow of feedwater to maintain the steam pressure. At the same time there was less cooling of the reactor due to the slowing turbine. This led to additional steam in the cooling channels, resulting in a surge of power 100 times the normal level.

The resulting sudden increase in temperature caused part of the fuel rods to rupture. Fuel particles reacted with the water causing a steam explosion that destroyed the reactor core. Another explosion followed two minutes later.

There were several reasons for this disaster:

- Those responsible for the Chernobyl Nuclear Power plant lacked a 'safety culture' resulting in an inability to remedy design weaknesses despite the fact that these were known about before the accident.
- The RBMK reactor type used at Chernobyl suffered from instability at low power and thus could experience a rapid, uncontrollable power increase. Although other reactor types have this problem they incorporate design features to stop instability from occurring. The cause of this instability is that water is a better coolant than steam, and that water acts as a moderator and neutron absorber (slowing down the reaction) whilst steam does not.
- Several safety procedures were ignored by the staff. For instance, despite a standard order stating that a minimum of thirty rods were required to retain

control, only six to eight rods were used during the test. The reactor's emergency cooling system was also disabled.

- The test was carried out without a proper exchange of information between the team in charge of the test and personnel responsible for the operation of the nuclear reactor.
- There was a long delay before those involved admitted the extent of the disaster either to themselves or to the public. Indeed, their behaviour in the first two days involved denial of the truth in an attempt to minimize the implications of the disaster.

The effects of the disaster have been calculated in various ways, often on the basis of inadequate information, once more showing that a lack of transparency leads to very different estimates. One report estimates that the amount of radioactivity released by the disaster was at least 200 times the amount released because of the Hiroshima and Nagasaki bombs (Schrader-Frechette 1999). Estimates of immediate deaths vary from 56 to just over 2500. Others estimate that it will result in almost half a million premature cancer deaths, and another half a million nonfatal cancers (Schrader-Frechette 1999).

After the Chernobyl accident, radioactive material was widely dispersed. It was measurable and showed effects over a vast area. The effects have been felt all over, practically the whole of the northern hemisphere. In some local ecosystems, lethal doses were reached particularly for coniferous trees and small mammals within a 10-km radius of the reactor. By 1989 the natural environment of these ecosystems began to recover, but there is the possibility of long-term genetic effects.

Schrader-Frechette (1999) suggests that the events of Chernobyl provide an example of environmental injustice, the idea that we have obligations to the environment in relation to the degree that we might affect it. Given the power of the reactor there was little attention given to how its failure might affect the environment.

Of course, it may not always be precisely clear as to what we do owe the environment. Hence, the Rio Declaration of 1992 set out the precautionary principle, stating that:

> In order to protect the environment, the precautionary approach shall be widely applied by States according to their capabilities, where there are threats of serious or irreversible damage. Lack of full scientific certainty shall not be used as a reason for postponing cost-effective measures to prevent environmental degradation.
>
> (Grace and Cohen 2005, 150)

This means that governments and business should anticipate unintended consequences and also the potential effect of any major disaster in making any major decision with respect to the environment. This is a higher standard than the law, placing the responsibility on those who engage in a potentially harmful activity. It looks to prove that something is not dangerous, the opposite of the final judgement in the Challenger case. It asks basic questions about what might go wrong and what measures are in place for that eventuality.

Anglo American provides a more effective example of responding to environmental challenges. The multinational corporation has signed up to the GRI report. The GRI involves triple bottom line reporting on the economic, environmental and societal dimensions of the corporation, underlining the point that all three are interconnected.

The environmental section includes information on:

- Total material use
- Direct energy use
- Indirect energy use
- Total water use
- Impacts on biodiversity
- Greenhouse gas emissions
- Ozone-depleting emissions
- Total amount of waste
- Environmental impact of products

Anglo American accepts that such reporting has been relatively recent for them and that there is a long way to go for their whole corporation. They report fines and legal actions taken against them (73 per cent down from 2004), and environmental incidents (level two incidents up by 5 per cent), alongside reference to awards and effective partnerships. On energy efficiency, the 2005 report gives a summary of work across their different companies. This includes an aim of 10 per cent reduction in carbon intensity over ten years. On air quality, sulphur dioxide emissions decreased by 43 per cent in one company. On water, there is sustained attempt to preserve fresh water and neutralize acidic waste water. The section on biodiversity lists work where companies have been involved in land stewardship and reclamation projects.

Reporting of this nature serves to establish benchmarks for performance, but also seeks to engage the imaginations of the different companies. A good example of this is the case study noted in the 2005 report in turning waste water from mines into drinking water.

Case 7.2

Drinking water for the Emalahleni Municipality

Years of mining in the area around Witbank in Mpumalanga, South Africa, have disrupted natural water cycles. Water that would otherwise flow into rivers is leaking into mines, where coal deposits make it acidic. This hampers mining activity and can lead to pollution of local water supplies. At the same time, growing demand from local communities and industry is draining supplies from local reservoirs.

'We saw an exciting opportunity to solve this problem by converting a mining environmental liability into a sustainable public/private partnership asset by addressing the water shortage challenge facing the local municipalities in the district', said hydrologist Peter Gunther.

Anglo American and project partner Ingwe began exploratory work in 2002 on the feasibility of a plant that would convert waste water from the mines to drinking water standards. Local communities and water regulators were closely involved in the plans and the project was given the go-ahead in 2005.

The plant and storage dams are being constructed at Anglo Coals Greenside colliery. The water treatment plant will neutralize acidic water from mines, remove metals and salt and chlorinate the water. Water quality will be monitored regularly.

Waste products from the treatment process will be disposed of alongside other waste from Greenside mine. Ways of recycling – and possibly selling – waste minerals such as limestone, magnesite and sulphur are being explored.

The plant, to be completed by 2007, will provide about 20 per cent of the Emalahleni municipality's daily water requirements. Local communities will also benefit from about 25 permanent positions at the plant and between 100 and 150 temporary jobs during construction.

Neighbouring municipalities are already planning to adopt a similar approach (Anglo American Report to Society 2005, 32).

Current progress in sustainable development has reached an important stage. Questions about how to make the project work, with cheap labour, resources and new markets as central issues are being replaced by questions about the purpose of the activity and how it affects the social and physical environment. In turn this has focused attention on the way in which major corporations function as never before. It is in the interest of such corporations to sign up for environmental reporting and development, not least to make sure that governments do not seek to ensure compliance. Questions will still remain about the justice of allowing activities such as mining in the first place and about what fair compensation might be for these.

Audits

The extent of environmental audits is taken further by the British Safety Council (Britishsafetycouncil.co.uk) in their Five Stage Environmental Sustainability Audit. This is divided into five levels:

Level 1. Environmental compliance. This looks to develop standard organizational compliance.

Level 2. Environmental management. This involves the development of core principles and benchmarks.

Level 3. Environmental management systems. This includes supply chain management.

Level 4. Environmental performance evaluation. This looks to develop environmental reporting, including the GRI, as part of an environmental risk strategy.

Level 5. Environmental sustainability. This includes developing greenhouse gas reduction policy, biodiversity action plan and ecological footprinting (monitoring the effects of the business on the environment).

Clearly, such an approach would take time to develop, culminating in the kind of proactive response shown by Interface (Case 7.3). The different levels resonate with Hunt and Auster's (1990) suggestion of the five stages of environmental commitment of any business. These involve:

Beginner. There is little resource commitment, and no management strategy.

Firefighter. Environmental issues are addressed when necessary, often only in response to legal requirements.

Concerned citizen. With a minimal budget, environmental issues are part of the management strategy.

Pragmatist. Environmental management is an accepted business function, and has a sufficient budget.

Proactivist. Environmental issues are a priority, with top management actively leading.

One of the most remarkable examples of a proactive company is Interface.

Case 7.3

Interface Inc

Interface deals with commercial and institutional interiors including carpeting. The head of Interface, Ray Anderson, began with looking at the reasons for getting involved.

Why is striving for sustainability so important?

Here's the problem in a nutshell. Industrialism developed in a different world from the one we live in today: fewer people, less material well-being, plentiful natural resources. What emerged was a highly productive, take–make–waste system that assumed infinite resources and infinite sinks for industrial wastes. Industry moves, mines, extracts, shovels, burns, wastes, pumps and disposes of four million pounds of material in order to provide one average, middle-class American family their needs for a year. Today, the rate of material throughput is endangering our prosperity, not enhancing it. At Interface, we recognize that we are part of the problem. We are analyzing all of our material flows to begin to address the task at hand.

What's the solution? We're not sure, but we have some ideas. We believe that there's a cure for resource waste that is profitable, creative and practical. We must create a company that addresses the needs of society and the environment by developing a system of industrial production that decreases our costs and dramatically reduces the burdens placed upon living systems. This also makes precious resources available for the billions of people who need more. What we call the next industrial revolution is a momentous shift in how we see the world, how we operate within it, what systems will prevail and which will not. At Interface, we are completely reimagining and redesigning everything we do, including the way we define our business. Our vision is to lead the way to the next industrial revolution by becoming the first sustainable corporation, and eventually a restorative enterprise. It's an extraordinarily ambitious endeavor; a mountain to climb that is higher than Everest (Interfaceinc.com).

Such a concern leads to the Interface Mission Statement.

Mission statement

Interface will become the first name in commercial and institutional interiors worldwide through its commitment to people, process, product, place and profits. We will strive to create an organization wherein all people are accorded unconditional respect and dignity; one that allows each person to continuously learn and develop. We will focus on product (which includes service) through constant emphasis on process quality and engineering, which we will combine with careful attention to our customers' needs so as always to deliver superior value to our customers, thereby maximizing all stakeholders' satisfaction. We will honor the places where we do business by endeavoring to become the first name in industrial ecology, a corporation that cherishes nature and restores the environment. Interface will lead by example and validate by results, including profits, leaving the world a better place than when we began, and we will be restorative through the power of our influence in the world.

This demanding mission statement leads to equally demanding objectives:

Objectives

1. Eliminate waste: Eliminating all forms of waste in every area of business.
2. Benign emissions: Eliminating toxic substances from products, vehicles and facilities.
3. Renewable energy: Operating facilities with renewable energy sources – solar, wind, landfill gas, biomass, geothermal, tidal and low impact/small-scale hydroelectric or non-petroleum-based hydrogen.
4. Closing the loop: Redesigning processes and products to close the technical loop using recovered and bio-based materials.
5. Resource-efficient transportation: Transporting people and products efficiently to reduce waste and emissions.
6. Sensitizing stakeholders: Creating a culture that integrates sustainability principles and improves people's lives and livelihoods.
7. Redesign commerce: Creating a new business model that demonstrates and supports the value of sustainability-based commerce.

The underlying perspective of this is two-fold. First, there is a concern with the way in which any product of the business affects the environment. Second, there is concern for how every aspect of the business, including every part of the product life cycle, can embody sustainable practice. Interface is quite clear that this does not simply mean reducing waste, protecting wildlife or recycling. These are important but only part of a process of continual reflection, which balances the development of the person with awareness of the environment and the capacity to respond to its needs.

Hence,

> Understanding and adopting sustainable business practices requires a new awareness of the world; the whole world, its natural systems and all of its species. It requires a deeper understanding of how the Earth works, and how man's processes affect nature's delicate balance. Understanding sustainability requires an awareness of how everything we do, everything we take, everything we make and everything we waste affects nature's balance, and how our actions will ultimately affect our children and the children of all species (Interfaceinc.com).

Underlying this are several core values, including:

- Respect for the members of the firm. This is based around a view of humanity as continuously learning and developing.
- Responsibility for the state of the environment along with other stakeholders.
- Commitment to find ways of responding to the environmental crisis.
- Concern to enable all stakeholders to develop an environmental awareness, and to give them opportunities for responding to environmental needs.
- Concern to play a part in restoring the environment. In this Interface lock in, consciously or not, to cultural values such as *kyosei* and *shalom*. The first is a Japanese communitarian concept meaning 'harmony' or 'working together for the common good'. It has been stressed by the work of the Caux Round Table, a group of major global industries working through CSR. The second is a Hebrew term meaning peace and justice. It stresses the need to restore relationships.

Interface sets out a holistic approach to environmental concern and sustainability, one that seeks to involve all stakeholders. In the first years of this approach it also increased profits.

Hence, the British Safety Council can argue that by focusing on such a process there are clear benefits to the company, including:

- Improved corporate reputation
- Increased sales to greener clients
- Improved productivity
- Increased profitability
- Improved staff morale

- Improved quality
- More control over organizational issues
- Better corporate risk management
- Improved relations with key stakeholders
- More environmentally sustainable ways of doing business (britsafe.org)

Summary and conclusion

The ethics of sustainable development suggest that a concerted effort must be made to promote the principles of sustainability at a local and community level. Sustainable development is connected with ideas from community development to produce an integrated process for securing sustainable communities (Didham 2002).

The concept of sustainable development presents a new set of ethical issues for the civil engineering profession. A unified approach to these issues is complicated by the lack of a standard definition, different practices domestically and overseas and a mixed bag of regulations from various levels and professional bodies within the construction industry provide confusing and conflicting definitions of sustainability.[4]

In response the engineering firm first needs to be aware of the environmental legal frameworks in whatever country it is operating. Second, it needs to work with stakeholders in achieving a shared view of sustainability and what is possible through creative partnerships.

This chapter has looked at the ethical issues related to sustainability, and how concern for this can be embodied in the firm. If these are followed integrally, then sustainable development will have a significant impact on the construction industry. This chapter also assessed issues associated with social justice, the triple bottom line of sustainable development and industry best practices of future sustainable development on the economic, environmental and social dimensions of the corporation.

Question

With two other students write an environmental policy for your company, imaginary or real, bringing together underlying values about the environment,

[4] http://construction-institute.org/xlc/week1.cfm (accessed on 27 February 2006).

the role of your business in relation to the environment, details of how awareness of good practice could be raised with employees and stakeholders and details of how practice is monitored.

References

Books

Arndt, H.W. (1993). Review article: sustainable development and the discount rate. *Economic Development and Cultural Change*, **41**, 3.

Bentham, J. (1897). *An Introduction to the Principles of Morals and Legislation*. Oxford: Clarendon Press.

Carson, R. (1962). *Silent Spring*. Boston: Houghton Mifflin.

Des Jardins, J.R. (2006). *Environmental Ethics – An Introduction to Environmental Philosophy*. Belmont: Thomson Wadsworth.

Grace, D. and Cohen, S. (2005). *Business Ethics*. Oxford: Oxford University Press.

Hunt, C. and Aster, R. (1990). Proactive environmental management. *Sloan Management Review*, Winter, 9.

Lovelock, J. (1979). *Gaia: A New Look at Life on Earth*. New York: Oxford University Press.

Robinson, S. (1992). *Serving Society: The Social Responsibility of Business*. Nottingham: Grove Ethics.

Schrader-Frechette, K. (1999). Chernobyl, global environmental justice and mutagenic threats. In *Global Ethics and Environment* (N. Low, ed.) pp. 70–89, London: Routledge.

Vesilind, P. and Gunn, A. (1998). *Engineering, Ethics and the Environment*. Cambridge: Cambridge University Press.

Warren, M.A. (1980). *Do Potential Persons Have Rights and Responsibilities to Future Generations*. Buffalo: Prometheus Books.

Wilson, E.O. (1989). Threats to biodiversity. *The Scientific American*, September issue, 60.

Reports

Anglo American (2005). *Report to Society*. London: Anglo American.

Barry, B. (1983). *Intergenerational Justice in Energy Policy*. MacLean and Brown.

Bruntland, G.H. (1987). *Our Common Future – The World Commission on Environment and Development*. Oxford: Oxford University Press.

Chew H.H. (1999). *Martin Buber's Philosophy of Dialogue as a Foundation for Environmental Ethics*. The National University of Singapore.

Didham, R. (2002). *The Case of Eco-citizenship – Shaping the Future through Sustainable Development and Community Development*. University of Edinburgh: The Centre for the study of Environmental Change and Sustainability.

Hurka, T. (1992). *Sustainable Development – What Do We Owe to Future Generations?* The Centre for Applied Ethics, University of British Columbia.

Leao-Aguiar, Ferreira and Marinho (2006). *Integrating CSR, Ethics and Sustainable Development Principles into the Construction Industry*. Federal University of the Bahia.

Lehni, M. (1992). *World Business Council for Sustainable Development – Business Strategies for Sustainable Development*. Deloitte & Touche.

Oyem, R.A. (2002). *Sustainable Ethics: The Divine Principle to Protect and Manage the Earth*. International Research in Geographical and Environmental Education.

8 *Global ethics*

Think globally, act locally
Friends of the Earth

Engineering is increasingly seen in a global context. In this chapter we will:

- Draw out and explain the phenomenon of globalization
- Examine ethics in a global context, with a particular emphasis on human rights
- Examine and analyse the global corporate social responsibility of a major engineering company using Anglo American as an example.
- Examine related issues such as the importance of working with NGOs and having a strategy for working across different cultures

Globalization

Globalization is a contested concept in social, economic or moral terms. Steger suggests this all-encompassing definition:

> Globalization refers to a multidimensional set of social processes that create, multiply, stretch and intensify worldwide social interdependencies and exchanges while at the same time fostering in people a growing awareness of deepening connections between the local and the distant.
>
> (Steger 2003)

- At the core of this process has been the expansion of capitalism across political boundaries. This has not simply been the intensification of worldwide trade – something evident since the fifteenth century. It also involves the organizing of production across national boundaries. This may be the result of critical resources or because of lower labour costs. This has led to the development of huge MNCs, some of which have annual financial turnovers larger than the gross national product of nations.
- Distance has been 'shortened' through rapid technological change, leading to significant increases in speed of communication and transport. The Internet

relays information in seconds and news-gathering technology can enable real time reporting of major events.

- All this has led to what Steger (2003, 11) calls the 'expansions and stretching of social relations, activities and interdependencies'. In retail, for instance, Western supermarkets are now stacked with produce from across the world. The justification for some of these relationships is questioned and this in turn has led to the development and proliferation of global NGOs such as Oxfam, Fair Trade, or Amnesty International.
- The 'opening up' of the world has led to increases both in migrations, with the workforce moving across national boundaries, and in cultural interaction. Business has had to learn how to operate in very different cultures. These changes have led, some would argue, to changes in national identity, accelerating the process of the breakdown of local communities (2003, 12).

Making sense of all this is different, and opinion tends to be divided into two broad views.

Globalization as good

Several arguments see globalization as having very good effects:

- One basis for this is the view that globalization involves freeing and integrating markets across the world. This is really a global market argument that suggests that the market is the best way of distributing goods between the nations, whilst at the same time ensuring that nations remain free.
- It is also argued that the globalization is now inevitable. This is partly due to the increased recognition that the different parts of the world are interdependent. This argument gives the impression of globalization as an unstoppable force. If that is so then no one is in charge of it, and the key to handling it is working together.
- In the light of the above it is argued that globalization benefits everyone, leading to greater efficiency, an increase in jobs and the equalizing of incomes between countries.
- It is argued that allied to an economic growth there is global development of shared values, not least through the development of democracy and human rights across the world.

Globalization as bad

There are broadly two groups that contest globalization. The first involves neoliberal groups in the United States that want to protect the freedom and culture of their own nation against the pervasive global influence. The second group focuses on the problems of the market approach and the need to develop ways to address global inequalities. This group argues that:

- The global market is of itself inequitable and needs political control. The idea that the market distributes equitably assumes that all in the market have similar power and similar opportunities, whereas, in reality, there are massive differences in power. The worldwide market, with monetary and trade systems, often disables the poorer countries, not least through massive interest on debt they have incurred.
- The global market is in fact dominated by the MNCs that are concerned primarily with their profits.
- There is little sense in the development of democracy. In any case why should democracy be prescribed for the rest of the world? This raises major questions about respect for the sovereignty of nations.
- Globalization does not respect the uniqueness of the different global cultures. Indeed, it leads to what has come to be known as McDonaldization – the domination of a consumer-led culture.

Not only, then, is the concept contested, but the effects of globalization and its ethical base are contested. Multinational engineering firms are very much part of this phenomenon, raising questions about how the ethics of working in such areas should be handled. One way of trying to deal with this is to develop a global ethics that can set universal standards by which the involvement of MNCs in the wider world can be judged.

There have been several attempts to develop such an approach for business. Some, such as the Caux Round Table, have focused on basic principles that cross cultural meaning. The group, including companies such as Canon and Philips, determined two basic guiding principles, human dignity and *kyosei* (working together for the common good or harmony). They charted these principles through four basic levels: economic survival; cooperation with the workforce; cooperating with stakeholders; global activism. The last of these argues the responsibility of MNCs to help governments redress core global imbalances, in the areas of:

- Trade
- Technology
- Wealth
- Environment

Other groups, such as the UN Global Compact, involve several thousand companies worldwide signing up to basic principles, focused in human rights.

The principles of the UN Global Compact

Human rights

Principle 1 Businesses should support and respect the protection of internationally proclaimed human rights.

Principle 2 Businesses should make sure that they are not complicit in human rights abuses.

Labour standards

Principle 3 Businesses should uphold the freedom of association and the effective recognition of the right to collective bargaining.

Principle 4 The elimination of all forms of forced and compulsory labour.

Principle 5 The effective abolition of child labour.

Principle 6 The elimination of discrimination in respect of employment and occupation.

Environment

Principle 7 Businesses should support a precautionary approach to environmental challenges.

Principle 8 Businesses should undertake initiatives to promote greater environmental responsibility.

Principle 9 Businesses should encourage the development and diffusion of environmentally friendly technologies.

Anti-corruption

Principle 10 Businesses should work against all forms of corruption, including extortion and bribery. *(unglobalcompact.org/TheTenPrinciples)*

Global ethics

Is it possible to have a global ethics – where certain principles are accepted universally? Hans Kung argues strongly for this, summarized in the stirring slogans,

> There will be no survival of our globe without a world ethic. No peace among the nations without peace among the religions. No peace among the religions without dialogue and cooperation among the religions and civilizations.
>
> (Hans Kung 1991, xv)

Globalization raises many issues of injustice and strife, from financial and labour markets to ecology and organized crime. Hence, such a global ethic is necessary if the global order is to be managed. As we have seen in the recent history of international terrorism, a great deal of the strife in this context comes from the interrelation of different religious groups with politics. Hence, argues Kung, there is a need to develop peace between religions. Such a peace would demand wide-ranging dialogue between religions, cultures and civilizations. The West tends to assume that society is secular and religions in the minority. Globally, it is quite the opposite; a majority of the world's population is strongly committed to a religious faith.

Kung argues that a global ethic is not a uniform system of ethics but rather 'a necessary minimum of shared ethical values' to which different regions, cultural groups, religions, nations and other interest groups can commit themselves. This involves an ongoing process of dialogue that uncovers the shared values already implicit in major ethical principles. The commandment 'Thou salt not kill', for instance, becomes, in positive terms, 'Have respect for life', calling for the safety of all minorities, social and political justice, a culture of non-violence, respect for the environment and universal disarmament. The commandment 'Thou shalt not steal' becomes 'Deal honestly and fairly', standing out against poverty and the cyclical violence that occurs in a society of wealth extremes. Just economic institutions need to be created and sanctioned at the highest levels, suggests Kung, and limitless consumption curbed in the developed countries while the market economy is made socially and ecologically conscious. 'Thou shalt not lie' becomes 'Speak and act truthfully', with particular focus on the media and politicians to become truly representative and accountable.

Principles and responsibilities that arise from this can be affirmed by all persons with ethical convictions, whether religiously grounded or not, who oppose all forms of inhumanity. Such an ethic is aspirational rather than prescriptive, forming the basis of the Declaration of Human Responsibilities (Kung and Schmidt 1998), intended as a complement to the UN Declaration of Human Rights.

Human rights

Of course, the Human Rights Declaration (see the end of this chapter for the full declaration) itself attempts to embody a universal ethic. It is made up of broad rights that are both negative, protecting certain 'goods', such as the right to life, liberty and property, and positive, seeking to enable certain goods, such as care for children and the provision of free education. This can be seen as a combination of prescriptive and aspirational principles, ones that should be followed now and others that a nation or intermediate organization would try to embody over time.

The preamble to the UN declaration of Human Rights sees such rights as derived from the 'inherent dignity of the human person'. This points to a prelegal and prepolitical moral belief in some key human characteristics that require universally specifiable ways of responding to human beings, and ways of prohibiting others. The rights should be protected by the rule of law.

Attempts to frame a universal ethic have been criticized on philosophical and social grounds. Firstly, the more specific the rule or right the more difficult it is to find universal support for it (Forrester 2005, O'Donovan 2000). Lewis (1978) argued for a universal ethic, based around a general acceptance of the various versions of the Golden Rule, 'Do unto others what you would have them do onto you'. This is such a general principle that it seems unexceptionable. More detail is needed to understand the ethical meaning in context, and the detail suggested by both Kung and the Human Rights declaration soon becomes contested. Human rights article 18, for example, suggests a right to change religion, but in some countries religion is tightly bound to national identity, and there are religious and legal constraints against this. Equally, rights surrounding freedom of speech or asylum seekers are subject to major questions in practice. What are the limits to freedom of speech and how is it possible to deal equitably with all who claim asylum?

General moral principles then seem unexceptionable but when faced by the particular context come up against the difficulty of how to embody such principles in practice. Underlying this difficulty is the phenomenon of moral pluralism, the vast array of different moral beliefs apparent in the world. Even within one nation, such as Britain, there has been a gradual breakdown of any meta-narrative, an overarching 'story' that gives shared spiritual or moral meaning (Connor 1989). This has been replaced by a form of liberal tolerance of a wide diversity of beliefs, principles and practice.

Secondly, philosophers such as Rorty (1993) argue that human rights attempt to provide a rational basis to ethical practice, and that ethics cannot be founded purely on that. Ethics demands an affective as much as a cognitive base, and this involves some degree of subjectivity. Bauman (1993) goes further to suggest that shared responsibility is the base of ethics and that this is pre-rational.

Thirdly, rights and even responsibilities assume the basis of a democratic nation, one that has the resources to deal with the various issues. Many developing nations do not have such resources. In any case, there are very different views of democracy (Chenyang 1999).

Fourthly, it is argued that the Declaration of Human Rights is Eurocentric, focused in a particular Western view of what is good. This contrasts sharply with other cultural views. The very term 'right' is legalistic and prescriptive, contrasting with Buddhist philosophy, for instance, which values mutuality, based on a view of interdependence, and underlying empathy. Such a philosophy does not have place for rights, looking rather to develop spiritual meaning as the base of ethics and responsive virtues.

Rights are based around individualistic views of equality and freedom. This contrasts, for instance, with the values of Confucianism. Whilst Confucianism is not totally averse to the idea of human rights (Twiss 1998), it stresses communitarian values, including paternalism and respect for authority, over freedom and equality. Others cultures, such as the Dinka in Sudan, have a strong sense of rights but only for the members of the community. Hence, these form the basis of conditional, not universal, rights (Drydyk 1999).

The dynamic of human rights is criticized as paternalistic. Mutua (2002), for instance, notes how the West characterizes the developing world as unable to develop ethical values without the help of the West, who are seen as 'saviours'.

Finally, Bauman (1989) argues that respect for ethical plurality is essential if the lessons of the Holocaust are to be learned. The Third Reich tried to impose moral meaning upon the world, one that ignored different cultures and the critiques that they might raise of any universal view of ethics.

The critiques of a universal ethic have some problems. Firstly, development of the Human Rights Declaration was decidedly not simply Western; there was global consultation (Twiss 1998). Hence some would argue that it is possible to have a global ethic that is not Eurocentric. Secondly, the examples of the Holocaust and subsequent genocides do point to the need to have some sense of justice and right that transcends national boundaries. This provides negative rational argument for the development of a shared ethic. As Nussbaum (1999) notes, without such a view, respect for different cultures can lead to support for oppression, reaction, sexism and racism. Thirdly, behind Kung's argument is an increasing sense of the world as interconnected, with nations mutually responsible for the shared physical and social environment and for long-term effects, such as global warming. Such potential disaster 'concentrates the mind wonderfully', as Charles I is said to have observed on his way to the scaffold, demanding shared responsibility and a shared ethic.

Hence, alongside respect for ethical plurality, there needs to be the balance of a shared ethic that transcends the particular interest of nations and major corporations. This leads inevitably to an ethics that is not easily or simplistically applied, requiring careful dialogue around the many different principles, cultural and religious perspectives, ideologies and social expectations that give meaning to any particular situation.

Child labour

A case that looks at differences at the heart of global values is that of child labour. On the face of it this would be a good candidate for a universal ethical rule: that child labour is wrong. This view led to vociferous arguments in the case of Nike. In the 1980s it was discovered that the labour force for some of Nike's shoes comprised mostly children working in Asian sweatshops. Nike have attempted, in response to activist groups, to address this, aware that condoning such practices could affect their Western sales. The debate is still very much alive (Murphy and Mathew 2001).

The argument against child labour is simply that it is a form of exploitation. Behind this are two factors. Firstly, children should be part of the education process. Secondly, children do not have any power and thus are unable to defend themselves against the exploitation. A further argument suggests that MNCs do not truly respect these children enough to pay a real wage.

Against this is the view that this is placing a Western perception on childhood that is not shared in a different culture. In a culture of poverty it may be that all members of the family have to earn money for the family to survive. Indeed, it may be that such a culture sees it as a matter of honour and respect for the child to participate in making money for the family. In such a culture the idea of full-time education may not be critical, desirable or even available.

It may be argued that the real difference here is not ethical. In both cases there is a concern to respect the child. The difference is rather in the socioeconomic context. But this only gives half the picture. Underlying this is the issue of the relationships between the wealthy developed world and the poor undeveloped world. In this case the developed world sees it as important to give opportunity to the child and has the means of ensuring that the average child in the developed world does have this right – a right that is enshrined in law. Philosophical rights then become legal rights that can be claimed. The non-developed world has neither the resources nor the legal framework necessary for universal education. The developed world also has the wealth and capacity to enable the non-developed world in this area. This raises the basic argument that if they can then they should help.

However, this brings us back to the question as to who will do the helping. Business cannot take full responsibility for such issues. It can accept, with governments, a shared responsibility, and then work through the different ways in which such responsibility can be fulfilled. At the basic level it is incumbent upon the MNC engineering firm to be aware of any human rights that may be violated in the supply chain. It is equally important to know if human rights are being violated by stakeholders, including any government that the company is working with. In addition, the aspirational sense of human rights gives any company a framework around which responsibility can be negotiated in context.

In light of such conflicting values, even if the principle against exploitation is seen as primary, there are real issues about how it might be embodied. Hence the Department for International Development stresses a complex debate, involving several questions:

- Should all forms of child labour or the worst forms be the focus of efforts?
- What is the impact of child labour on adult employment and wages?
- Are work and school incompatible in the lives of children?
- What are the most effective strategies against harmful forms of child work, e.g. labour legislation, compulsory education, poverty reduction, social mobilization?
- What are the pros and cons of consumer boycotts of or other sanctions against the products of child labour?
- Should children participate in decision-making processes, and if so how?
- What are the key alliances needed for effective policy implementation?
- What are the links between poor health, mortality and child labour – particularly in relation to HIV/AIDS (DFID 2001).

Such a debate includes a concern for basic principles, such as the autonomy of the children, how different options will affect the different stakeholders, how the issue of child labour links to wider problems such as health and poverty and how stakeholders, including governments, MNCs, NGOs and intermediate organizations, might work together to take account of the different values and constraints. Human rights ideology has to take account of these different, often competing, values.

Questions

Read through the Declaration of Human Rights at the end of this chapter.

How many of these rights would apply across all nations?

What are the core human rights?

Whose responsibility is it to uphold human rights?

What might a company do if a government in relation to the company's work were seen to be abusing human rights? (See resettlement below)

A good illustration of working out responsibility in a global context is the work of the Anglo American corporation.

Anglo American plc.

A core business of Anglo American is mining engineering. This takes the company into situations of environmental, cultural and political sensitivity, demanding a carefully worked out set of principles and policies.

The principles

Anglo American state in the preamble to their principles that they see the objectives of providing the best return for shareholders and fulfilling social and environmental responsibilities as complementary. A key ethical/economic concept used is that of stewardship. Stewardship has a strong religious history around the idea that humanity is made the stewards of creation (Robinson 1992). They are responsible for maintaining the integrity of the social and physical environment. Anglo American (2004, 1) signal that they have a part in the role of stewardship. This does not deny self-interest, as they write 'our operations will perform better when the communities surrounding them are stable and prosperous'.

Two other factors stand out in the preamble. The first point stresses that modern business is operating in the context of continuous change, which tends, at best, to be confusing. The default position then for most people is to fear the 'perceived motivations and power of international corporations'. The preamble suggests then that countering such a perception requires a great deal of work on the part of an MNC. At least this involves transparency and accountability on the part of the corporations with a clear indication of how the power and influence can be used for the good in society. The second point is really about the integrity of the corporation. Whilst Anglo American operates in many different countries, with different values and cultures, the values and principles of the corporation have 'universal application'. This in turn means ensuring that these principles are made clear and that they apply to every business managed by Anglo American.

It is now worth quoting noting these principles in full.

Good citizenship: our business principles

Our responsibilities to our stakeholders
Our primary responsibility is to our investors. We will seek to maximise shareholder value over time. We believe that this is best achieved through an intelligent regard for the interests of other stakeholders including our employees, the communities associated with our operations, our customers and business partners. A reputation for integrity and responsible behaviour will underpin our commercial performance through motivating employees and building trust and goodwill in the wider world.

The following considerations guide our dealings with stakeholders:
Investors
We will ensure full compliance with relevant laws and rules. We are committed to good corporate governance, transparency and fair dealing.

Employees
We aim to attract and retain the services of the most appropriately skilled individuals. We are committed to treating employees at all levels with respect and consideration, to investing in their development and to ensuring that their careers are not constrained by discrimination or other arbitrary barriers to advancement. We will seek to maintain a regular two-way flow of information with employees to maximise their identification with, and ability to contribute to, our business.

Communities
We aim to promote strong relationships with, and enhance the capacities of, the communities of which we are a part. We will seek regular engagement about issues which may affect them. Our support for community projects will reflect the priorities of local people, sustainability and cost effectiveness. We will increasingly seek to assess the contribution our operations make to local social and economic development and to report upon it.

Customers and business partners
We seek mutually beneficial long-term relationships with our customers, business partners, contractors and suppliers based on fair and ethical practices.

Governmental bodies
We respect the laws of host countries whilst seeking to observe, within our operations, the universal standards promulgated by leading intergovernmental organisations. We aim to be seen as socially responsible and an investor of choice.

Non-governmental organisations
We aim for constructive relations with relevant non-governmental organisations. Their input may lead to better practices and increase our understanding of our host communities.

Principles of conduct
Business integrity and ethics
We support free enterprise as the system best able to contribute to the economic welfare of society as well as to promote individual liberty. Without profits and a strong financial foundation it would not be possible to fulfil our responsibilities to shareholders, employees, society and to those with whom we do business.

However, our investment criteria are not solely economic. They also take into account social, environmental and political considerations. We will comply with all laws and regulations applicable to our businesses and to our relationships with our stakeholders. We are implacably opposed to corruption. We will not offer, pay or accept bribes or condone anti-competitive practices in our dealings in the marketplace and will not tolerate any such activity by our employees. We prohibit employees from trading securities illegally when in possession of unpublished price-sensitive information. We require our employees to perform their duties conscientiously, honestly and with due regard for the avoidance of conflicts between any personal financial or commercial interests and their responsibilities to their employer.

We will maintain high standards of planning and control to identify and monitor material risks; safeguard our assets; and to detect and prevent fraud. We will promote the application of our principles by those with whom we do business. Their willingness to accept these principles will be an important factor in our decisions to enter into, and remain in, such relationships.

We encourage employees to take personal responsibility for ensuring that our conduct complies with our principles. No one will suffer for raising with management violations of this policy or any legal or ethical concern.

Corporate citizenship

We respect human dignity and the rights of individuals and of the communities associated with our operations. We seek to make a contribution to the economic, social and educational well-being of these communities, including through local business development and providing opportunities for workers from disadvantaged backgrounds.

We recognise the sensitivities involved in addressing issues which relate to the cultural heritage of indigenous communities. We will seek to ensure that such matters are handled in a spirit of respect, trust and dialogue. We believe we have the right and the responsibility to make our positions known to governments on any matters which affect our employees, shareholders, customers or the communities associated with our operations. Whilst the primary responsibility for the protection of human rights lies with governments and international organizations, where it is within our power to do so, we will seek to promote the observance of human rights in the countries where we operate. We support the principles set forth in the Universal Declaration of Human Rights.

Employment and labour rights

We are committed to the adoption of fair labour practices at our workplaces and our conditions of service will comply with applicable laws and industry standards.

We will promote workplace equality and will seek to eliminate all forms of unfair discrimination. We will not tolerate inhumane treatment of employees including any form of forced labour, physical punishment, or other abuse. We prohibit the use of child labour. We recognise the right of our employees to freedom of association.

We will operate fair and appropriate means for the determination of terms of conditions of employment. We will provide appropriate procedures for the protection of workplace rights and our employees' interests. We will provide employees with opportunities to enhance their skills and capabilities, enabling them to develop fulfilling careers and to maximise their contribution to our business.

Safety, health and environmental stewardship

We have adopted a comprehensive Safety, Health and Environment Policy and will report regularly on our SHE performance. We will continue to review and develop this policy. We strive to prevent fatalities, work-related injuries and health impairment of employees and contractors. We recognise the need for environmental stewardship to minimise consumption of natural resources and waste generation and to minimise the impact of our operations on the environment. Senior executives and line management are accountable for safety, health and environmental issues and for the allocation of adequate financial and human resources within their operations to address these matters. We will work to keep health, safety and environmental matters at the forefront of workplace concerns and will report on progress against our policies and objectives. We recognise the human tragedy caused by the HIV/AIDS epidemic, particularly in sub-Saharan Africa. We have a clear policy for addressing HIV/AIDS in the workplace and are committed to a comprehensive prevention strategy, linked to programmes of care for those with HIV/AIDS. We will strive to eliminate any stigma or unfair discrimination on the basis of real or perceived HIV status. We are committed to the principles of sustainable development, by which we mean striking an optimal balance between economic, environmental and social development. We will strive to innovate and adopt best practice, wherever we operate, working in consultation with stakeholders.

This statement should be read in conjunction with fuller policy statements such as our Safety, Health and Environment Policy and such other codes and guidance notes which may be issued from time to time.

(Anglo American plc. 2004).

The Anglo American approach sets out what it means by ethics and what the basis of that ethics is. They identify the core purpose of responsibility to the shareholder, and argue that the interest of the shareholders is best served through 'intelligent regard' for other stakeholders. What that means will be spelled out in practice. Broadly, however, they believe that their primary aim cannot be achieved without some concern for the well-being of their employees and for trust and goodwill 'in the wide world'. So the basis of their ethical concern is mutual interest with some degree of mutual responsibility.

The document carefully distinguishes between responsibilities and principles, and these give more meaning to the initial paragraph. Several key factors stand out:

- There is a strong sense of mutuality in working with employees and community. The company seeks to identify with the local community and enable employees identify with the company.
- Mutuality involves respect for the rights of both persons and communities involved in the company operations. It extends to a concern for how the company can contribute to the well-being of the community. The company also signals its concern for genuine dialogue that respects the belief system of indigenous communities.
- This raises key issues of how the company relates to governments. In this, Anglo American are not simply aiming to abide by the local laws but also to uphold more universal standards. This places Anglo American in the position where they might have to confront current government practices, communicating to them human rights violations.
- Alongside any concerns for the dialogue and the development of awareness the company sets down some very clear principles, both negative (anti-corruption, child labour, unfair discrimination) and positive (the right of employees to freedom of association).
- The principles are also relevant to business partners. The intention to promote them in those partnerships is signalled. Indeed, this could affect which partnerships are developed.
- Much of the document is based in self- and mutual interest. However, it is clear that respect for human rights goes beyond that and there is also a broader concern for the well-being of social and physical environments. This is partly expressed in terms of health, safety and environmental issues.

Two issues sum up these concerns: HIV/AIDS and resettlement policies.

HIV/AIDS

This is, of course, the most significant health issue in sub-Saharan Africa, with inevitable consequences on any workforce. Self-interest dictates that any company provides the best possible care for their workers, and Anglo American had by 2004 provided anti-retroviral therapy for almost 25 000 of them. This is not simply a case of providing therapy. It is estimated that at least another 5000 employees may have the illness but have not come forward for treatment. Hence, there is need to work on raising awareness of the medical resources and working through why so many have not taken advantage of the care. Ethically, it could be argued that whilst this demonstrates respect and care for the employee not only is it in the interest of the firm but it might lead to real inequity in the community. Employees and their families have care available, whilst their neighbours may not. This in turn raises issues about the negotiation of responsibility. It could be argued that the local government or health authority should be the key responsible agent, not Anglo American.

The response of Anglo American is interesting. They have negotiated a Community HIV/AIDS Partnership Programme (Anglo American 2004). This aims to deepen and broaden the effect of the work-based approach focusing on young people and education. It is part of a behaviour change campaign that also looks to increase the numbers that take voluntary counselling and testing. The hope is to link it to current initiatives from other companies. Part of the project will involve close links to the relevant NGOs (see below). This is an ongoing project that involves more than simply self-interest and that through the wide partnerships can monitor the difference that is made. This is moving towards the company seeing itself as a stakeholder in the community.

Resettlement

The difficult side of working around this area of engineering emerges fully in the issue of resettlement. In a Friends of the Earth (2003 http://www.foe.co.uk/campaigns/corporates/case_studies/index.html) briefing it was noted that Anglo American partly owned a mine in Colombia. In order to develop the mine the local community had to be resettled. The briefing alleges that in 2002 the community around Tabaco was forcibly resettled by government paramilitaries, leading to 95 per cent unemployment. There was no proper compensation and Anglo American said nothing about the invasion of human rights.

The issue of resettlement is very difficult, not least because the data are not always confirmed. Anglo American counter the allegation with the view that the paramilitaries involved were not from the government, hence it was difficult for them to address the issue. There are other cases where there have been allegations of collusion, such as the resettlement of Sarahita. Anglo American note that there had not been a 'community' there for over a decade. What occurred was the eviction of squatters. The company also noted other cases where the local firm had been involved in helping massacre survivors (2004, 42). The point of raising these issues is not to argue for one side or the other but rather to illustrate the importance in ethical terms both of being concerned about issues such as resettlement and of getting the data right. The first of these recognizes a responsibility. The second can be very difficult, especially in a global context. Distances involved, different cultural perceptions and the part played by mass media can mean that data are distorted at a very early stage, getting locking into a contested dynamic that is hard to get out of. It therefore becomes critical that companies and stakeholders work together, and find ways of verifying the truth. The verifying of truth also links to transparency about the purpose and values of companies and other stakeholders. This dynamic is perhaps best illustrated by the issue of working with NGOs.

Working with NGOs

As noted above, NGOs have proliferated in the last decades of the twentieth century in response to the increase in globalization. There are arguments against NGOs.

- They are not fully accountable.
- They hold a brief that speaks for the oppressed and it is not clear what the democratic basis of this is.
- Their stance on justice is often around black and white, single issues. This often leads to a presumption about the values of business, with the NGO taking the stance of protecting weak groups against the might of the MNC. In some cases this has led to decisions not to speak to MNCs on principle.
- In pursuit of their perceived view of justice their methods can be very questionable.

However, NGOs are a very powerful constituency that can also highlight issues that need to be dealt with, and a good example of the need to work with them is the Brent Spar case study.

Case 8.1

Brent Spar

The Brent Spar was an offshore storage facility owned by Shell and based in the North Sea. In 1992, this huge facility, with six large storage tanks beneath the water, was due to be decommissioned. Because of its size, Shell argued against the usual practice of taking a structure into shore to have it broken up on land. Sinking in deep water, they argued, was in this case the cheapest way of dealing with the problem, and also, some claimed, the best environmentally. In discussing it with the government, agreement was reached with a date in 1995 established for the sinking of the facility. Shell had not consulted Greenpeace in this process.

On the 30 April 1995 Greenpeace occupied the Brent Spar, accompanied by German and UK journalists. Greenpeace reported that they had not been consulted and that the Spar was a 'toxic time bomb', with oil residues and radioactive waste that could seriously damage the marine environment. Dismantling would have to be on shore. In the subsequent battle, Greenpeace managed the publicity such that when the Spar began to be towed off to the deeper sea it appeared that there was a battle of integrity between profit-hungry MNCs and the NGO that represented the environment. The result was that Shell backed down under the pressure of intense publicity.

It was only later that the Greenpeace action and the data that they based it on were questioned. Shell argued that there was only 53 tonnes of toxic sludge or oil on the Spar. Greenpeace argued that there was over 145 000 tonnes of toxic rubbish and over 100 tonnes of toxic sludge. An independent study later noted that there was between 74 and 100 tonnes of oil on board the Spar, and that the greater part of this could be removed easily. The subsequent expert testimony that sinking the Spar in deep water was the least bad environmental option led Greenpeace to offer a public apology (for more details see Entine 2002)

Several critical issues emerged from this case, summed up by Jon Entine (2002):

- The contested use of science: risk–benefit analysis/cost–benefit analysis, 'junk science', environmental politics
- The role of the media
- Stakeholder reputation management and greenwashing by corporations and activist groups
- The problematic role of dialogue in stakeholder conflict resolution

In the first of these issues the very fact that Greenpeace was not consulted led to different perceptions of the situation. In turn this led to contested scientific judgement and the lack of clarity about the actual facts. It would have been in Shell's interest to open up the conversation at the earliest stages so that data collection could have been agreed. This in turn depends upon the development of dialogue between NGOs and major industry. However, given the often adversarial role of the NGOs, dialogue itself can become difficult. Some NGOs claim that if they are even seen to be in sustained conversation with global business, this will be perceived as 'selling out' by their supporters. Some means of ensuring the development of trust is therefore vital, not least in allowing NGOs to develop purpose that is not primarily adversarial.

If the relationship with NGOs is poor then it can become even worse through media involvement. The impetus to sensationalize tends to promote polarization and partiality and in this case there was the assumption in most of the press that big corporations lie and that activist groups pursue the truth. Hence, despite the more complex nature of the case, Greenpeace won the battle for public support in some way (Entine, 73). Nonetheless, the long-term reputations of Shell, Greenpeace and the UK government were all tarnished and took some time to recover. Perhaps the most striking aspect of this case was the awareness that activist and industry alike were using spin doctors in the attempt to win the battle.

Elkington (1997) suggested three conclusions from this case:

- Long-term life cycle planning was needed for major technological undertakings. In this case there should have been built-in decommissioning plans that could have acted as the basis for decision-making.
- Simple permission from regulators is no longer sufficient for such decisions.
- There should be transparency of process and efforts to achieve dialogue between business and the NGOs.

At the very least it is clear that the power exercised by many NGOs can be very effective and thus it is in the interest of the company to not have an adversarial relationship with them.

Underlying all of this is an important point for all global business. The very fact that there is such a disparity of power between the corporation and the poor nations and the communities in which they operate means that the default perception of most people is that the company is exploiting both the land and the people. Aside from any ethical questions it is hard not to assume that this is a basically unfair relationship. It is therefore in the interest of the company to carefully work against that perception.

There is a more profound underlying ethical point behind that. It is clear that MNC involvement in the developed world is on the basis of a global market system that is inequitable. The corporation *does* have immense power and sufficient resources to make a huge difference to the communities and environments in which they operate. Failure to use their resources to address some of these issues is failure to genuinely respect those communities, and can be seen as a form of exploitation. Justice requires that this be attended to. Note here the idea of justice as not simply equal treatment for same groups but also as based in need. Corporations such as Anglo American are developing in that direction.

Dealing with different cultures

Religion

In a global context, religion can play an important part of any business relationship (Vesilind and Gunn 1998, 108–24). This requires some key steps:

- It is important to have local knowledge advice. Religion is not monochrome. As already noted, for instance, there are many different kinds of Muslim groups, including over 2 million Muslims in Turkey who are largely pacifist.
- There is a need to be aware of sensitive theological areas that would affect any aspects of business relationships, not least in relation to respecting bans on alcohol or female dress codes. Another example of the need for sensitivity was the worldwide Muslim reaction to the printing of cartoons of the prophet Mohammad in a Danish newspaper. Western values accept that both politics and religion can be the object of satire. For many religions, however, satirical treatment of core beliefs is seen as blasphemy, and therefore a direct attack on the religion itself.
- There may be different underlying views about work or humanity. The value of autonomy, for instance, is perceived in the West as uniquely important for ethics. However, autonomy can easily be seen in some religions as giving too much weight to the individual. For some faiths this leads to a view of society and community that is fragmented and self centred. Such religions see the importance of community, in the sense of solidarity, and believe that obedience is more important than choice. This can in turn lead to reinforcement of values that can be inegalitarian, not least on gender issues. Again, this may conflict with Western views of equality as a core value.

The response to all this plurality includes the need to:

- Listen carefully for the very different values that may be embodied in practice.
- Focus on areas of shared practice-centred concern. This can build up trust and also tease out both shared and possibly conflicting values.
- Work on possible partnerships. Again, this can build up trust, not least through demonstration of commitment.

Cultures

A good example of different cultural approaches is bribery and corruption. Most firms have a statement that forbids the taking of bribes. Bribery is defined as any situation where there is an attempt to procure services outside the normal contract negotiations or any attempt to use public money for private gain. This is reinforced by La Federation Internationale des Ingenieurs Conseil (FIDIC). Their statement on this argues that bribery and corruption are both morally and economically damaging and therefore should be banned globally. If this were taken seriously, however, then it would be hard to see how business can be done in some countries.

There are in fact three traditional approaches to this issue:

- *Ethical conventionalism*. This argues that the conventions of the country should be followed. In some countries bribery is seen as a type of fee for work done.
- *Ethical fundamentalism*. This argues that the same high ethical standards be applied across different cultures.
- *Ethical casework*. In this, each case is examined separately.

The most effective way to operate would seem to be to have a clear aim in the firm's code of ethics to avoid bribery and corruption and then discuss how this applied on a case-by-case basis. Such a policy would require the firm to be clear about what is defined as bribery, and about accepted company reactions to any attempted bribe. It is relatively easy, for instance, to have the rule that no employee should accept a gift of more than £25. As Weiss (2003, 295) notes, this will be all the more effective if the ethical perspective is shared by several different companies, groups and nations.

Clarifying the situation may, of course, involve careful work with stakeholders, including government departments. In this case it is important to know the national policy on bribery, so that any attempts at bribery can be challenged in the light of that. The UK Anti-Corruption Forum (neill.stansbury@transparency.org.uk) stresses the importance of building up and becoming part of networks of governments, banks, professional associations and companies that work together to respond to this problem. Their newsletter no. 2 (May 2006) provides a summary of many different initiatives. However, this is a long-term response to bribery and corruption. There will always be cases where it is not clear if a bribe has been given – the ski trip offered as part of hospitality and so on. In all cases the definition of bribery has to be returned to undisclosed payments that are intended to influence the judgement of someone who has the power to decide in favour of the person offering the bribe. The ski trip may or may not be an explicit bribe, but when in doubt it should be disclosed, again taking the ethical issue away from simply an individual ethical dilemma to one that is responded to in the light of a transparent ethical culture.

Weiss (2003, 330) sums up an attempt to work across cultures with the idea of ethical navigation. This involves:

- Articulate basic choices in the light of company' and stakeholders' values.
- Be aware of the limits of responsible corporate power.
- Maintain responsibility in 'flexible business relationships'. This includes developing a reflective company such that the experiences of fellow professionals can be shared, allowing the development of a 'repertoire' of different approaches that embody responsibility.
- Negotiate amongst the different cultural values, whilst retaining one's own.
- Listen to and assert the different interests of the parties involved.

Weiss stresses the importance of getting the dynamic right, suggesting that conflict resolution techniques are important in working through any cross-cultural issues. Buller et al. (1991) suggest conflict resolution techniques including:

- *Avoidance*. This is a judgement as to when not to engage. If one party is in a stronger position than another then avoidance can help to avoid escalation of ethical conflicts.
- *Education/persuasion*. This focuses on sharing values. MNCs, for instance, have focused on the values of technology for host communities, but have also learned from host communities about the core values of sharing resources, community building and economic development. Such values are all too often seen as exclusive, but close listening and reflection can explicitly demonstrate how they can work together.
- *Infiltration*. This develops on the previous approach, allowing different groups to get used to values of the others.
- *Negotiation/compromise*. This focuses on responsibility and how all parties can contribute to a response.
- *Accommodation*. In this either party may find it useful to take on the values of the other, showing genuine openness.
- *Collaboration*. This is the final stage in embodying the sharing of responsibility.

In the case of bribery, such an approach could involve a company having a clear policy on its employees not accepting bribes. This would be a clear signal about the company's values. Secondly, the company could become part of, or develop, an anti-bribery network. Thirdly, it could look to ways of developing transparency, ensuring that any requests for bribes are reported. None of this can provide easy solutions where bribery and corruption are endemic.

Case 8.2

You are in charge of an overseas office of a UK engineering company. In the United Kingdom the company operates a strict policy of not accepting bribes. However, in the environment in which you work bribery is commonplace and accepted. Part of a major project is being delayed by equipment held by local customs officers, who are waiting for their usual 'payment' to release it. What would your policy be for this?

Suppose that on seeking advice from your UK office you are told to use your own initiative to expedite the situation. However, if you decide to make the payments you are told to 'hide' them in the accounts submitted for the UK company audit. How would you feel about this advice and what action would you take? What ethical principle would you base that action on?

Politics, war and ethics

The dynamics since the attack on the Twin Towers have heightened the awareness of global ethical issues. The presence of Western troops in Iraq has led to further reflection on the so-called 'warrior values' and the values of the 'occupying powers'. This in turn has even led to a focus on ethics training for the armed forces. Equally, it has led to significant questioning of political and company motives, not least with the association of the Bush government with Halliburton. This underlines the political context of much of CSR and global ethics, and the need to develop dialogue with governments.

Conclusions

In the context of globalization, firms have to work harder at being ethical and being seen to be ethical. They are faced by:

- NGOs that demand transparency.
- Media that regularly monitors business practice.
- Global organizations that are looking to develop broader ethical perspectives.
- Different cultures and religions that if not respected can cause major problems.
- Governments that may have very different and varied agendas.
- Major environmental issues shared across the globe.

In the light of all this the global perspective demands that companies have in place carefully considered ethical principles and policy. Such policy has to centre on how the ethical meaning can be developed through working closely with the other stakeholders involved.

This dialogue has to take seriously both a global ethical perspective, as summed up by human rights, and the particular and possible different values of different communities. If either is given too much precedence this can lead to injustice. Hence, the dialogue has to be ongoing.

Questions

1. Working with other students sketch out a policy for CSR for an engineering company working in Iraq.
2. Find a statement of bribery from three different MNCs on the World Wide Web. Do they show any differences in values of practice? Which one would you choose and why?
3. Working with other students work out government guidelines on the ethics of companies working in a war zone.

References

Anglo American (2004). *Report to Society*. London: Anglo American.

Bauman, Z. (1989). *Modernity and the Holocaust*. London: Polity.

Bauman, Z. (1993). *Postmodern Ethics*. Oxford: Blackwell.

Buller, K., Kohls, J. and Anderson, K. (1991). The challenges of global ethics. *Journal of Business Ethics*, 10, 767–75.

Chenyang Li (1999). Confucian value and democratic value. In *Moral Issues in Global Perspective* (C. Koggel, ed.) pp. 74–81, Peterborough, Ontario: Broadview.

Connor, S. (1989). *The Post Modern Culture*. Oxford: Blackwell.

DFID [Department for International Development] (2001). *Child Labour*, Key Sheet.

Drydyk, J. (1999). Globalization and human rights. In *Moral Issues in Global Perspective* (C. Koggel, ed.) pp. 30–42, Peterborough, Ontario: Broadview.

Elkington, J. (1997). Laying the ghost of Brent Spar. *Resurgence*, 3, September.

Entine, J. (2002). Shell, Greenpeace and Brent Spar: the politics of dialogue. In *Case Histories in Business Ethics* (C. Megone and S. Robinson, eds) pp. 59–95, London: Routledge.

Forrester, D. (2005). *Apocalypse Now?* Aldershot: Ashgate.

Kung, H. (1991). *Global Responsibility: In Search of a New World Ethics*. London: SCM.

Kung, H. and Schmidt, H. (1998). *A Global Ethic and Global Responsibilities*. London: SCM.

Lewis, C.S. (1978). *Abolition of Man*. New York: Prentice Hall.

Murphy, D. and Mathew, D. (2001). *Nike and Global Labour Practices*. A case study prepared for the New Academy of Business Innovation Network for Socially Responsible Business.

Mutua, M. (2002). *Human Rights: A Political and Cultural Critique*. Philadelphia: University of Philadelphia Press.

Nussbaum, M. (1999). Human functioning and social justice: in defence of Aristotelian essentialism. In *Moral Issues in Global Perspective* (C. Koggel, ed.) pp. 124–46, Peterborough, Ontario: Broadview.

O'Donovan, O. (2000). Review of a global ethic and global responsibilities. *Studies in Christian Ethics*, 13(1), 122–28.

Rorty, R. (1993). Human rights, rationality, and sentimentality. In *On Human Rights* (S. Shute and S. Hurely, eds.) pp. 101–32. New York: Basic Books.

Robinson, S. (1992). *Serving Society: The Responsibility of Business*. Nottingham: Grove Ethics.

Steger, M. (2003). *Globalization*. Oxford: Oxford University Press.

Twiss, S. (1998). Religion and human rights: a comparative perspective. In *Exploration in Global Ethics* (S. Twiss and B. Grelle, eds) pp. 155–75, Oxford: Westview.

Vesilind, P. and Gunn, A. (1998). *Engineering, Ethics and the Environment*. Cambridge: Cambridge University Press.

Weiss, J. (2003). *Business Ethics*. Mason: Thomson.

Box 8.1

The Declaration of Human Rights

On 10 December 1948, the General Assembly of the United Nations adopted and proclaimed the Universal Declaration of Human Rights, the full text of which appears in the following pages. Following this historic act the Assembly called upon all member countries to publicize the text of the declaration and 'to cause it to be disseminated, displayed, read and expounded principally in schools and other educational institutions, without distinction based on the political status of countries or territories.'

Preamble

- Whereas recognition of the inherent dignity and of the equal and inalienable rights of all members of the human family is the foundation of freedom, justice and peace in the world,
- Whereas disregard and contempt for human rights have resulted in barbarous acts which have outraged the conscience of mankind, and the advent of a world in which human beings shall enjoy freedom of speech and belief and freedom from fear and want has been proclaimed as the highest aspiration of the common people,
- Whereas it is essential, if man is not to be compelled to have recourse, as a last resort, to rebellion against tyranny and oppression, that human rights should be protected by the rule of law,

- Whereas it is essential to promote the development of friendly relations between nations,
- Whereas the peoples of the United Nations have in the Charter reaffirmed their faith in fundamental human rights, in the dignity and worth of the human person and in the equal rights of men and women and have determined to promote social progress and better standards of life in larger freedom,
- Whereas Member States have pledged themselves to achieve, in co-operation with the United Nations, the promotion of universal respect for and observance of human rights and fundamental freedoms,
- Whereas a common understanding of these rights and freedoms is of the greatest importance for the full realization of this pledge,

Now, Therefore THE GENERAL ASSEMBLY proclaims THIS UNIVERSAL DECLARATION OF HUMAN RIGHTS as a common standard of achievement for all peoples and all nations, to the end that every individual and every organ of society, keeping this Declaration constantly in mind, shall strive by teaching and education to promote respect for these rights and freedoms and by progressive measures, national and international, to secure their universal and effective recognition and observance, both among the peoples of Member States themselves and among the peoples of territories under their jurisdiction.

Article 1.
All human beings are born free and equal in dignity and rights. They are endowed with reason and conscience and should act towards one another in a spirit of brotherhood.

Article 2.
Everyone is entitled to all the rights and freedoms set forth in this Declaration, without distinction of any kind, such as race, colour, sex, language, religion, political or other opinion, national or social origin, property, birth or other status. Furthermore, no distinction shall be made on the basis of the political, jurisdictional or international status of the country or territory to which a person belongs, whether it be independent, trust, non-self-governing or under any other limitation of sovereignty.

Article 3.
Everyone has the right to life, liberty and security of person.

Article 4.
No one shall be held in slavery or servitude; slavery and the slave trade shall be prohibited in all their forms.

Article 5.

No one shall be subjected to torture or to cruel, inhuman or degrading treatment or punishment.

Article 6.

Everyone has the right to recognition everywhere as a person before the law.

Article 7.

All are equal before the law and are entitled without any discrimination to equal protection of the law. All are entitled to equal protection against any discrimination in violation of this Declaration and against any incitement to such discrimination.

Article 8.

Everyone has the right to an effective remedy by the competent national tribunals for acts violating the fundamental rights granted him by the constitution or by law.

Article 9.

No one shall be subjected to arbitrary arrest, detention or exile.

Article 10.

Everyone is entitled in full equality to a fair and public hearing by an independent and impartial tribunal, in the determination of his rights and obligations and of any criminal charge against him.

Article 11.

1. Everyone charged with a penal offence has the right to be presumed innocent until proved guilty according to law in a public trial at which he has had all the guarantees necessary for his defence.
2. No one shall be held guilty of any penal offence on account of any act or omission which did not constitute a penal offence, under national or international law, at the time when it was committed. Nor shall a heavier penalty be imposed than the one that was applicable at the time the penal offence was committed.

Article 12.

No one shall be subjected to arbitrary interference with his privacy, family, home or correspondence, nor to attacks upon his honour and reputation. Everyone has the right to the protection of the law against such interference or attacks.

Article 13.

1. Everyone has the right to freedom of movement and residence within the borders of each state.

2. Everyone has the right to leave any country, including his own, and to return to his country.

Article 14.

1. Everyone has the right to seek and to enjoy in other countries asylum from persecution.

2. This right may not be invoked in the case of prosecutions genuinely arising from non-political crimes or from acts contrary to the purposes and principles of the United Nations.

Article 15.

1. Everyone has the right to a nationality.

2. No one shall be arbitrarily deprived of his nationality nor denied the right to change his nationality.

Article 16.

1. Men and women of full age, without any limitation due to race, nationality or religion, have the right to marry and to found a family. They are entitled to equal rights as to marriage, during marriage and at its dissolution.

2. Marriage shall be entered into only with the free and full consent of the intending spouses.

3. The family is the natural and fundamental group unit of society and is entitled to protection by society and the State.

Article 17.

1. Everyone has the right to own property alone as well as in association with others.

2. No one shall be arbitrarily deprived of his property.

Article 18.

Everyone has the right to freedom of thought, conscience and religion; this right includes freedom to change his religion or belief, and freedom, either alone or in community with others and in public or private, to manifest his religion or belief in teaching, practice, worship and observance.

Article 19.

Everyone has the right to freedom of opinion and expression; this right includes freedom to hold opinions without interference and to seek, receive and impart information and ideas through any media and regardless of frontiers.

Article 20.

1. Everyone has the right to freedom of peaceful assembly and association.
2. No one may be compelled to belong to an association.

Article 21.

1. Everyone has the right to take part in the government of his country, directly or through freely chosen representatives.
2. Everyone has the right of equal access to public service in his country.
3. The will of the people shall be the basis of the authority of government; this will shall be expressed in periodic and genuine elections which shall be by universal and equal suffrage and shall be held by secret vote or by equivalent free voting procedures.

Article 22.

Everyone, as a member of society, has the right to social security and is entitled to realization, through national effort and international co-operation and in accordance with the organization and resources of each State, of the economic, social and cultural rights indispensable for his dignity and the free development of his personality.

Article 23.

1. Everyone has the right to work, to free choice of employment, to just and favourable conditions of work and to protection against unemployment.
2. Everyone, without any discrimination, has the right to equal pay for equal work.
3. Everyone who works has the right to just and favourable remuneration ensuring for himself and his family an existence worthy of human dignity, and supplemented, if necessary, by other means of social protection.
4. Everyone has the right to form and to join trade unions for the protection of his interests.

Article 24.

Everyone has the right to rest and leisure, including reasonable limitation of working hours and periodic holidays with pay.

Article 25.

1. Everyone has the right to a standard of living adequate for the health and well-being of himself and of his family, including food, clothing, housing and medical care and necessary social services, and the right to security

in the event of unemployment, sickness, disability, widowhood, old age or other lack of livelihood in circumstances beyond his control.

2. Motherhood and childhood are entitled to special care and assistance. All children, whether born in or out of wedlock, shall enjoy the same social protection.

Article 26.

1. Everyone has the right to education. Education shall be free, at least in the elementary and fundamental stages. Elementary education shall be compulsory. Technical and professional education shall be made generally available and higher education shall be equally accessible to all on the basis of merit.
2. Education shall be directed to the full development of the human personality and to the strengthening of respect for human rights and fundamental freedoms. It shall promote understanding, tolerance and friendship among all nations, racial or religious groups, and shall further the activities of the United Nations for the maintenance of peace.
3. Parents have a prior right to choose the kind of education that shall be given to their children.

Article 27.

1. Everyone has the right to freely participate in the cultural life of the community, to enjoy the arts and to share in scientific advancement and its benefits.
2. Everyone has the right to the protection of the moral and material interests resulting from any scientific, literary or artistic production of which he is the author.

Article 28.

Everyone is entitled to a social and international order in which the rights and freedoms set forth in this Declaration can be fully realized.

Article 29.

1. Everyone has duties to the community in which alone the free and full development of his personality is possible.
2. In the exercise of his rights and freedoms, everyone shall be subject only to such limitations as are determined by law solely for the purpose of securing due recognition and respect for the rights and freedoms of others and of meeting the just requirements of morality, public order and the general welfare in a democratic society.

3. These rights and freedoms may in no case be exercised contrary to the purposes and principles of the United Nations.

Article 30.

Nothing in this Declaration may be interpreted as implying for any State, group or person any right to engage in any activity or to perform any act aimed at the destruction of any of the rights and freedoms set forth herein.

9 *Call yourself an engineer?*

Isambard Kingdom Brunel was one of the great icons of engineering. There was about the man an extraordinary competence, confidence and congruence.

Born in 1806 – his father an innovative engineer in his own right – by the age of twenty he was the engineer in charge of the difficult works involved in constructing the Thames tunnel, overcoming disasters and displaying the outstanding qualities that would exemplify his working career. His imagination knew few bounds, moving with ease from tunnels to bridges, railways to harbours, to prefabricated buildings and even ships. He crossed the divides between the civil engineer, mechanical engineer, naval architect, and architect with ease, always at the cutting edge of the current technology of his age, taking his collaborators with him to the limits of their capacities.

As an entrepreneur he anticipated and exploited developing markets throughout his life. It cannot be said that he made his fortune, or fortunes for his backers, but his contribution to society was such that he can be spoken of as one the great Britons (Cooper 2002). Like his fellow Victorian engineers, there was a sense in which all the different aspects that we have written about in this book came together in this man. He lived in a time of settled values and a confidence in progress but the changes in engineering and technology following on from the start of the industrial revolution heralded changes in society and the professions that served them. In Brunel's lifetime the professions of mechanical and civil engineering were separated with the foundation of the Institution of Mechanical Engineers. A process of ever-increasing specialization began, which appears to grow faster in our lives today as technical and scientific knowledge increases so rapidly so to make specialization inevitable. The sum of knowledge today has become too enormous for any one intellect, however brilliant, to contain.

The age when men like Brunel could practice so competently in so many areas of expertise is over. The deaths of the three great Victorian engineers, Brunel, Stephenson and Locke, in 1859, all three dying prematurely within a few months of each other, marked the end of an era. Each seemed to have hurried to fit an extraordinary lifetime of work into an impossibly short period.

Brunel was not only a great engineer but also a great man; his forceful personality and professional philosophy, his imaginative power and his wide-ranging liberal intellect made him great. In his speech for the first Brunel Lecture in 1958 Rolt wrote,

> One thing that strikes one immediately about him is his insistence throughout his life upon his absolute personal responsibility for any work which he undertook. He was often asked to act as a consulting engineer but always refused. In one such letter of refusal he summed up his attitude in these words: The term 'Consulting Engineer' is a very vague one and in practice has been too much used to mean a man who for a consideration sells his name but nothing more. Now I never connect myself with an engineering work except as the Directing Engineer who, under the Directors, has the sole responsibility and control of the engineering and is therefore 'The Engineer'. In a railway, the only works to be constructed are engineering works, and there can really be only one engineer.

Certainly, there was only one engineer of the Great Western Railway – no one connected with the Company from the directors down to the navvies was left in any doubt on that score. In these days of committees, consultants and specialist subcontractors it may seem almost incredible that a young man of thirty should have absolute responsibility for every detail of the building of a railway from London to Bristol, yet so it was.

Today it is very different. With the complexity of business and society all these things have become fragmented: enterprise, management, engineering and values. And with this fragmentation can easily come the erosion of meaning, the polarizing of competence and ethical values and the denial of responsibility. It has been the theme of this book that the engineer can bring these things together once more. It is no longer possible to bring them together in one person. It is possible to do so if engineering as a whole can have a big enough shared vision of its underlying values, its place in global society and its responsibility to that society. In this book we have tried to show how all this can be developed and held together, including:

- The ethical identity of the engineer, and the underlying values and virtues.
- The method of ethical reflection that can include values in everyday decision-making.
- The methods of institutional reflection and monitoring that can bridge values, practice and identity, in professional codes and in business accounting.
- The ways in which responsibility can be negotiated, not denied.

None of this suggests that any professional or company can be perfectly ethical. On the contrary, we suggest that ethics is really about a learning process, and thus about accepting limitations, developing transparency and learning from reflection on practice.

In this chapter we aim to draw together some of these issues in a broader context. The chapter will examine:

- Some ideas about the philosophy of engineering.
- The debate about technology and society.
- The case of nuclear energy as an example of this debate
- How engineers might respond to some of these broader issues.

The philosophy of engineering

One of the most positive views of the engineer comes from Samuel Florman. Florman (1976) suggests that engineering is in effect a very high calling, which involves fundamental 'existential pleasures'. Existential means experiential, involving the whole person – reason, feeling and physical. Engineering, he suggests, is an attempt to engage with and utilize the social and physical environment in order to fulfil human needs, desires and aspirations. This involves several existential pleasures:

- The very act of being able to change the world in some way. There is a human impulse to change and improve, and the pull of these endless possibilities 'bewitches the engineer of every era'.
- The joy of the applied scientist who is able to begin to understand the laws of the universe in the context of the creative enterprise. This is not a sterile or simply functional relationship to the universe, or a grasp of numbers and formulae. Florman suggests that it is a relationship to the environment that can actually involve 'quasi-mystical moments of peace and wonder' (141).
- The engineer is also involved in response to what Florman calls 'mammoth undertakings' that appeal to the human passions (122).
- Florman also suggests that the engineer finds pleasure through using technology. This is partly the pleasure of control and of attempts to solve problems.
- There is finally the pleasure of service. Florman writes, 'The main existential pleasure of the engineer will always be to contribute to the well-being of his fellow man' (147). Florman calls on the testimony of engineers who have reflected on their work (94 ff.). What makes the task worthwhile is its contribution

to improving human life. There may be debate as to what that means, as we shall see below, but the idea of contribution to society remains constant.

Such a philosophy holds together several things:

- Awareness of the physical and social environment along with the particular skills of the engineer. Carter (1985) reinforces this with the stress that competence does not have to close off the mind and the spirit from the wider world. Hence, the skills of engineering go hand in hand with the virtues.
- The idea of service with the idea of enterprise. Enterprise is often thought of as being simply about making money, and by extension as something that conflicts with ethics. Florman's view enables us to see enterprise as something that is creative and exciting in itself and also as something that always has a social implication. In philosophy and theology this brings together the two ideas of agape (unconditional care for others) and eros (care that is conditioned by the attraction of the other). All too often these have been polarized (Robinson 2001). The engineer, along with the other professionals, can, and should, have both a care for society and a care about the creation of the project.
- The idea of achievement, and creation, with the idea of uncertainty. Martin and Schinzinger (1989) add to this the view that the task of the engineer is fundamentally one of an experimenter. Whilst technical calculation may give the impression of certainty they argue firstly that any engineering project is carried out in partial ignorance, involving uncertainties in the abstract model of design calculation, the qualities of material utilized and the loads to which the finished product will be subjected. Secondly, the side effects of any engineering project are uncertain. The dam that brings water to communities may eventually degrade the local ecosystem. Hence, there is need to monitor progress after the project is finished to review unintended consequences and learn from this. If the engineer is essentially an experimenter then this reinforces the view that the creative process itself involves risk. Risk has appeared throughout this book: the risk at the heart of the Challenger case; the risk of taking a business decision that might affect the reputation of the company; the risks covered in any project audit. A major task of the engineer then is to weigh risks and benefits and report to those involved in the whole process so as to determine what is an acceptable level of risk. A bad example of this was the infamous Ford Pinto case (Cullen 1987). This car had a major design defect, with the petrol tank in an exposed position at the rear, making it highly vulnerable to rear-impact collisions. The design fault was pointed out to the company, but they determined they it would cost the company less if they did not modify the vehicle construction

assembly line, and thus initially decided to accept the risks to life involved. The calculations were as follows:

> **Ford Pinto**
>
> **Benefits of altering the design**
> Savings: 180 deaths; 180 serious injuries; 2100 vehicles
> Unit cost: \$200 000 per death; \$67 000 per serious injury; \$700 per vehicle
> Total benefit: \$49.5 million
>
> **Costs of altering the design**
> Sales: \$11 million cars; 1.5 million trucks
> Unit cost: \$11 per car; \$11 per truck
> Total cost: \$137.5 million

Such calculations remind us of the importance of seeing cost–benefit and risk–benefit analysis in a broader ethical framework.

It is not always possible to remove all risk from an engineering project, and often there is a point where any attempt to reduce risk completely could make a project no longer viable. But what is an acceptable level of risk? Each project requires that risk be calculated, and perhaps the most important part of this process is that it should be transparent, so that all stakeholders are aware of the possible risks and who takes responsibility for them. In the Challenger case the process was fragmented, so that the launch controller was not aware of the risks. In the Pinto case the calculations were 'top secret'. In the case of nuclear power, as we shall see below, it becomes very hard to calculate risks effectively. Moreover, the calculation of risks in the nuclear context makes us re-examine the underlying ethical foundation of the argument for increased nuclear power. It is presumed that we should simply accept the demand for power at its present or higher levels, when perhaps this situation should be questioned and consumers encouraged to take more responsibility for reducing power demand.

Technology and society

The engineer is part of the world of technology and thus part of the constant debate about the place technology has in society. Peter Davies (2001) suggests that this debate can be seen in terms of three heads: technology as 'just more of the same', with the same problems relative to different times; technology as neutral and technology as progress.

The relativity of technology

Writers such as Wenk (1986) see modern technology as relatively the same as previous technology. From the wheel to the spinning jenny, new technologies have startled society and caused them to fear the consequences to livelihood and ethical meaning. Even the first heart transplant was seen by many as a shocking arrogation of God's role. Twenty-first century technology involves simply more changes, in travel, communication to which we have to adapt, and gradually will. Jonas (1984), however, argues that modern technology is different from previous different moral challenges:

- Technology involves an inherent ambiguity. The aim of much technology is good, but often has unintended consequences down the line. DDT, for instance, was developed with a good end in view, and seemed to work well before the dangerous side effects were discovered. This argument can be taken further along the lines that we can never fully know what the consequences of technology are.
- The global scale and power of technology. Technological decisions affect all aspects of the social and physical environment. Such power brings huge responsibilities. Charting the effect of that power and how responsibility is affected becomes ever more important.
- The apocalyptic potential of technology. Technological advances now give humanity the power to destroy both itself and its environment. Hence, the development of technology shows a real difference from the past. The prime example of this is nuclear power (see below).

Technological determinism

A further key element in Jonas's argument is the view that once technological developments are under way they are unstoppable. This easily slides into an imperative that if something can be done then it should be.

Ellul (1964, 17) goes further, arguing that 'Modern technology has become a total phenomenon for civilization, the defining force or a new social order in which efficiency is no longer an option but a necessity imposed on all human activity'. In this situation technology itself is the base of our values and beliefs. In effect we hold it sacred, an end in itself. For Ellul this means that technology has become in one sense autonomous, by which he means not controlled by society. This, he argues, reduces the human being to a 'slug in a slot machine' (Ellul 1964, 135). Bauman gives a different perspective from this, arguing that technology's aim is to make things easier and that this inevitably leads to techniques such as the division of labour. The division of labour itself leads to a fragmentation of community and breakdown of communication and identity. So-called experts rule over this process, but they are not focused on core values, such as community and respect, but on the process itself. Hence, the process is not value driven and can become prey to pathological views such as those seen in the Third Reich. We come once more then to Willie Just and the gas trucks on the Eastern Front. He was simply a cog in the wheel of the Third Reich, which used technology to its ends. Behind that, argues Ellul, is the only principle of the technological society, that is 'efficient ordering'.

Such technology inevitably begins to shape society. We perceive life in the light of technology. Problems are set out as technical or technological ones and thus technology is seen as the only way of solving them – the so-called technological fix. Such a view is effectively used to create modern 'needs', which are, themselves, consumer 'inventions', such as the need for new cars. A good example of this dynamic is the world of medical technology. Since the Second World War the development of medical technology has been enormous. It has been perceived that improved technology would have the capacity to solve all the health problems. This has resulted in the specifying of targets, such as reducing waiting lists. Whilst these can be seen as good ends, they have become ends that dominate the service to the exclusion of other issues – not least the development of health education. Conditions such as obesity, diabetes and heart disease are on the increase and cannot be dealt with through technological fixes, such as drugs or operations, but through developing a healthy life style. Illich (1977) takes this argument further in relation to medical technology in criticizing the use of technology by doctors. He argues that there are three major problems that this had created:

- Use of medical technology has led patients to not take responsibility for their lifestyle.
- Medical technology, including drugs, has led to a massive increase in iatrogenic illness, illness caused by doctors, not least through the side effects of drugs.
- Doctors have come to rely increasingly on technical tests for diagnosis rather than on their differential diagnostic skills. This has led to further increases in technology.

Technology as neutral

The simple argument here is that technology is a tool, as such it can be used to good and bad ends. The problem then is not the technology itself but the people and nations who use the technology. However, this apparently simple distinction becomes less clear when the interaction of wider society is considered. Technology, like any human action, embodies underlying values. The choice of one technology over another also embodies the interests of the group that chooses. Sustainable technology such as wind and wave power has constantly lost out to other technologies precisely because of a choice by governments, which is based upon the perceived need to maintain and possibly increase the energy consumption levels of a nation. The assumed values are materialist and consumerist. Often, they will be present without any explicit articulation of them.

Postman (1993) goes further in terms of information technology. He argues that uses of technology are 'determined by the structure of the technology itself' (1993, 7). He argues that technology such as the computer and the television are not simply machines that convey information. They are also ways in which we conceptualize reality. We see the world through them and believe how they classify it, value it and frame it.

Davies (2001) also suggests that the idea of technology as neutral can lull us into an easy acceptance of the underlying values and into seeing life in terms of technological problems with technological solutions. The complexity of how technology relates to the different power structures can thus easily be lost, and with it our capacity to challenge technological developments in a constructive way.

Technology as progress

Since the industrial revolution there has been a strong sense that technological innovation was itself enabling 'progress'. Such progress was almost self-evident – an increase in productivity and the creation and distribution of wealth, important both for individuals and for the nation. Indeed, the technological wonders created at that time could make people of the United Kingdom justly proud to be world leaders. Progress in all this is very much a value word – something about the good consequences of the innovations, and something about the pride in the skills, and even about identity. Hence, opposition to such progress was seen in a negative light and the 'Luddites' are often perceived as self-evidently against progress, thus obstructing what is good of the majority.

Davies (2001) notes three factors that have led to what is in effect an ideology of progress:

1. The term 'progress' is loaded. Technological progress has become associated with material progress and with social and moral progress.
2. History, not least because it is most often written by the winners, assumes a gradual betterment. Hence, we see the different, and improving ages: the Stone Age, the Bronze Age and the Iron Age. The industrial revolution has moved into the age of globalization, followed by the era of computing. Greater delights are created to surprise us and improve our well-being, and the presumption is one of progress.
3. Intellectual and technological skills and achievements are very quickly woven into a broader view of moral progress. Technology leads to the development

of ways in which connections can be made with the world, building bridges literally and metaphorically between communities, and with that, so the argument goes, comes peace and civilizing values. Of course this is nothing new. As early as the fourth century Augustine was suggesting, along with Irenaeus, that completely spiritual and material advancement would lead to a golden age. Later Theologians such as T. P. Forsyth could chart how, with the beginning of the twentieth century, a new golden age was dawning. Real signs of moral progress were argued for. Then came the Great War. Technologies that were previously at the service of war-like masters returned to their original role, and carnage reached new heights. Even this could not totally quench a sense of resilience, with post-war efforts to harness technology to the principle of equality culminating with the founding blocks of the Welfare State.

The picture then is not a simple one of technology as assisting progress or even of technology as progress. Technology ushers in progress but the progress brings with it still critical questions about purpose and meaning. Technology can make work safer and improve a target-based approach, but without attention to the meaning of that work a fundamental dimension can be lost. As noted above, the health service is a good example of this, with targets such as cutting waiting lists threatening to take attention away from the fundamental values of patient care (Robinson et al. 2003). Another important example of technological advance is the Internet. This has revolutionized communication. However, like all advances, it carries with it problems, including:

- The sheer size of the networks and the subsequent overload of information and communication encountered by users.
- The psychological distance of communication that can lead to insensitivity.
- The global nature of cyberspace, with no common laws or codes.

The Internet is not a value-free zone; it is part of a much broader civil society and ethical sense needs to be made of it. This may mean developing codes of practice.

Question

One way of handling the problems of the Internet is to devise a code of conduct for Internet users. Working with a fellow student, suggest ten basic rules for such a code.

The power of the idea of 'technology as progress' lies essentially in its materialist values. Technological progress does not of itself lead to reflection about how such goods are produced or distributed, the effects of such goods on different groups or the nature and purpose of any progress. Davies suggests that our underlying debates about technology are not as sophisticated as they might be. It is much easier to see what technology frees us from – activities, including work that is not fulfilling or mundane. It is less clear what technology frees us for, or how it connects to a wider-meaning system.

Perhaps the ultimate irony is that we now have a technology with a global reach operating in a global community that includes massive inequity and poverty. In one sense this brings the debate back to Aristotle and the need to reflect regularly on the underlying purpose of any technology. Simply to assume that technology is good because it makes life more convenient or takes away the hassle of cooking is not clear. What is the view of the good life behind that? Evidence would suggest that fast foods have contributed to an increase in obesity and diabetes, causing the meaning of progress to be questioned in this context.

Finally, whilst it would be wrong to see technology as totally deterministic it is true that it is filled with unintended consequences. The classic case is the automobile. Who could have foreseen that when cars were first invented that they would give rise to two massive industries – petrochemicals and cars? Both industries have given the world new freedoms. But who could have foreseen that the world would become dependent upon oil? Who could have foreseen that the petrochemical industry could become a crucial counter for national interests, or that wars could be fought around oil? Who could have foreseen the effects of the car: the accident death toll, the way in which it has changed lifestyles, the effect of exhaust pollution on the ozone layer? At no point in that history has there been one person or group to blame for these effects. These are simply negative effects associated with the use of cars. The important point is that no one could have known what the results would be.

In one sense this can be seen simply as a comment on the limitations of humanity. If this is so then it raises the question of how we ever can control technology. Joseph Rotblat (2001, 175 ff.), who worked on the Manhattan project, in his acceptance speech of the Nobel Peace Prize argued that the scientists and engineers who worked on the first nuclear bomb could not have known how this work would lead to the proliferation of such weapons. Nonetheless, he argues that they were responsible. He quotes Lord Zuckerman, the then Chief Scientific Adviser of the British Government,

> When it comes to nuclear weapons . . . it is the man or woman in the laboratory who at the start proposes that for this or that arcane reason it would be useful to improve and old or to devise a new nuclear warhead. It is them, the technicians, not the commander in the field, who are at the heart of the arms race (176).

Of course the technician was not *solely* responsible. However, applied scientists shared responsibility, and in some cases encouraged the army and politicians to see further possibilities. Again, this brings us back to some of the key points about ethics in the Chapters 1 and 2. The scientists and engineers in the Manhattan project did not ask the basic questions about what it all meant.

This issue is alive for us today in the context of debate about whether greater use would be made of nuclear power, ironically, to meet targets of sustainability.

Nuclear power

After several decades of distrust of nuclear power it is now back on the agenda. Two things have caused the rethink. Firstly, there is increasing energy consumption throughout the world, whist traditional sources of energy are becoming rarer. Secondly, nuclear energy seems to be the most environmentally friendly option.

Some arguments for nuclear power

Sustainability. A nuclear reactor emits virtually no carbon dioxide, compared to the major greenhouse gas emitted by other types of power generation.

Safety. Nuclear power is relatively safe. There have been only three major incidents, the Windscale fire of 1957, Three Mile Island in 1979 and Chernobyl 1986. These failures included an old Soviet design, which, before some very extensive modifications to the type, precipitated the Chernobyl disaster. Over 10 000 reactor-years of operation have shown a remarkable lack of problems in any of the reactors that are licensable throughout the world. There is probably no other large-scale technology used worldwide with a comparable safety record. This is largely because safety was given a very high priority from the outset of the civil nuclear energy programme, at least in the West. About one-third of the cost of a typical reactor results from its safety systems and structures, including containment and back-up provisions. This is a higher proportion of cost than even aircraft design and construction. Any statistics comparing the safety of nuclear energy with alternative means of generating electricity show nuclear to be the safest. The oil and coal industry have caused far more deaths over time and the safety lessons have been learned. Potential risk is reduced further by the greater inherent safety of modern nuclear reactors.

Economy. Modern nuclear reactors are far more economic than older versions. Uranium prices have remained secure for some time, giving greater fiscal security than given by fossil fuels.

Energy security. From a national perspective, the security of future energy supplies is a major factor in assessing their sustainability. Whenever objective assessment is made of national or regional energy policies, security is a priority.

France's decision in 1974 to dramatically expand its use of nuclear energy was driven primarily by considerations of energy security. However, the economic virtues have since become more prominent. The EU Green Paper on energy security in 2000 put forward coal, nuclear energy and renewable as three pillars of future energy security for Europe. The US government is clear that nuclear energy must play an increasing role this century.

Opportunity costs. Nuclear power gives access to virtually limitless resources of energy with negligible opportunity cost, i.e. what other options might be negatively affected. Like renewables, such as wind or solar power, nuclear power does not deplete resources useful for other purposes, using energy that is relatively abundant. A mixed approach to energy could then begin to meet the demands of future generations.

Waste. Nuclear power produces wastes in operation and decommissioning. However, these are two relatively safe options and the nuclear industry is the only one to cost these into the product.

Geological disposal of nuclear waste is the most widely accepted as an option:

> The scientific and technical community generally feels confident that there already exist technical solutions to the spent fuel and nuclear waste conditioning and disposal question. This is a consequence of many years of work by numerous professionals in institutions around the world. . . . There is a wide consensus on the safety and benefits of geological disposal (OECD 2001).

Underlying ethics. Pescatore (1999) suggests four basic principles. These were generated through discussions with the US government about principles for guiding decisions by public administrators on the basis of the international Rio and UNESCO declarations. Both of these acknowledged responsibilities to future generations. The principles are:

- The Trustee Principle: 'Every generation has obligations as trustee to protect the interests of future generations.'
- The Sustainability Principle: 'No generation should deprive future generation of the opportunity for a quality of life comparable to its own.'
- The Chain of Obligation Principle: 'Each generation's primary obligation is to provide for the needs of the living and succeeding generations,' the emphasis being that 'near-term concrete hazards have priority over long-term hypothetical hazards'.
- The Precautionary Principle: 'Actions that pose a realistic threat of irreversible harm or catastrophic consequences should not be pursued unless there is some countervailing need to benefit either current or future generations.'

Applying these principles to the question of nuclear waste, and in particular to their geological disposal, a system with inherent passive safety, Pescatore (1999) notes:

- The generation producing the waste is responsible for its safe management and associated costs.
- There is an obligation to protect individuals and the environment both now and in the future.
- There is no moral basis for discounting future health and risks of environmental damage.
- Our descendants should not knowingly be exposed to risks that we would not accept today. Individuals should be protected at least as well as they are at present.
- The safety and security of repositories should not presume a stable social structure for the indefinite future or continued technological progress.
- Wastes should be processed so that it is not a burden to future generations. However, we should not unnecessarily limit the capability of future generations to assume management control, including possible recovery of the wastes.
- We are responsible for passing on to future generations our knowledge concerning the risks related to waste.
- There should be enough flexibility in the disposal procedures to allow alternative choices. In particular, information should be made available so the public can take part in the decision-making process, which, in this case, will proceed in stages.

Pescatore points out that geological disposal is considered as the final stage in waste management. He argues that it ensures security and safety in a way that will not require surveillance, maintenance or institutional control.

The argument for the development of nuclear power ends by stressing that thus far there have been no credible alternatives set out for future large-scale production of electricity. Solar and wind capacity cannot begin to answer the needs on this scale. In the light of that, and of the equally important concern for sustainability, nuclear power should play a prominent part in a mixed economy of power.

Some arguments against nuclear power

Against this view is that of the Sustainable Development Commission in the United Kingdom (2006). In the context of government targets of a 20 per cent cut in carbon dioxide emission by 2010 and 60 per cent cut by 2050, the Commission argues that even if nuclear power was double that in the United Kingdom this would not achieve an 8 per cent cut in carbon dioxide emission by 2035. The development of the energy source would take too long.

In addition, there are several major problems with nuclear power:

- There are no long-term solutions yet for dealing with nuclear waste. Because of the nature of the waste, it carries massive risks. Deep dumping of the waste was rejected by a UK public enquiry in 1997 precisely because the science of dealing with such waste remains unproven. There can be no guarantee of isolating the waste from the environment. Whilst ten new reactors would result in only 10 per cent increase in radioactive waste they would also result in a 300 per cent increase in the most long-lived wastes. The precautionary principle should, in the light of these considerations, cause us to hold back on the nuclear option, rather than move forward.
- The ultimate cost of the nuclear option remains uncertain. Once built, they will produce low-cost power, but the capital cost, of construction and decommissioning, is very hard to estimate before they are built.
- At a point when there is unprecedented interest in networks that will enable local power distribution, the focus on nuclear power would press Britain into a centralized power system.
- Nuclear power would give consumer and business the impression that the real problems are technological and that this is the technological fix. This would take the debate away from questions about needs and responsibilities of consumers and the need for energy efficiency. This would be a good example of technological determinism.
- If Britain develops nuclear power it is difficult to see how the West could deny other nations, such as Iran and North Korea, the opportunity to do the same, especially in the light of climate change negotiations. It is then highly possible that developing nations could have lower safety standards, thereby creating higher risks of radiation exposure and terrorist attacks.
- Despite high safety standards in the West, power plants would still be possible targets for terrorists.
- Nuclear power is often perceived as a domestic source of energy. In fact the United Kingdom has no uranium deposits, and this raises serious questions about the long-term availability and the short-term supply of uranium.
- The Commission argued that there could be significant opportunity costs, not least the strong possibility that, with the need to focus quickly on developing the nuclear option, this work could take away resources from the promotion of energy efficiency.

In a commentary on the Commission's view Jonathan Porritt (2006) argues that there is a wider ethical consideration that takes us back to sustainability itself. The Bruntland report defined sustainability as 'development that meets the needs of the present without compromising the ability of future generations to meet their own need' (1987). This can be seen as maintaining intergenerational equity. Porritt continues,

> Nuclear technologies pose complex ethical dilemmas in this regard. High-level nuclear waste remains dangerously radioactive for hundreds of thousands of years; nuclear reactors will have to be 'moth balled' for decades whilst decommissioning takes place. A portion of the risks and of the ongoing costs associated with waste management and decommissioning will therefore fall on citizens who are neither party to the decision taken to build the reactors in the first place, nor beneficiaries of the electricity that flowed from those reactors during their life time.
>
> (Porritt 2006, 13)

The nuclear debate effectively shows that there are no easy answers. At one level, nuclear power provides real value conflict, between the capacity of this power to reduce carbon emissions and the possible future effects of nuclear waste and decommissioning. It also highlights the virtual impossibility of anticipating the consequences of any one option. Finally, it reminds us of another major argument, pinpointed by Davies, as to whether we need more or less technology. The move forward in technology would seem to question the very nature of our humanity. A recent news headline noted the choice, in a medical context, to abort foetuses with club feet. Technology allows us to see the problems and to 'deal' with them. Underpinning such a view of technology are two related views: the idea that human beings can themselves be 'improved' if not made perfect and the idea that risk can be completely eliminated. The danger of the first is that it can lead eventually to eugenics. The danger of the second is that it can lead to a lack of awareness of the realities of the social and physical environment.

This is central to the nuclear power issue demanding that certain key principles, suggested by the Sustainable Development Commission (2006, 3), guide the debate:

- Responsible use of science. This requires both rigorous evidence for any decision and acceptance of scientific uncertainty.
- Promoting good governance. This involves both effective governance and decision-making that enables broad debate and public engagement. One of the dangers of the present situation is that major players with resources can begin to dominate the debate. The nuclear lobby, for instance, has become associated with campaigns against wind and sun power. Government has to ensure transparency and space to hear all views and fears.
- Achieving a sustainable economy.

- Ensuring a strong, healthy and just society. This involves meeting the different needs of all the people. Of all the principles suggested by the Commission this one comes closest to developing a positive underlying ethic or purpose, rather than seeing technology as simply freeing society from constraints.
- Living within environmental limits. This involves respecting the limits of the environment and biosphere. It asks major questions at all levels about what energy is needed to sustain humanity. In turn this looks to cut back on demand. Inevitably, this is connected to questions about purpose. It also moves any debate away from simple rights to responsibility. In this case responsibility is shared by people at all levels of society.

The engineer and technology

How then should the engineer respond to this debate? Davies suggests that this is not something that should be left to philosophers or theologians. They clearly have opinions, and thus should be part of the interdisciplinary debate. However, by very dint of their role, engineers 'have a responsibility to articulate in both word and deed, and ethical position in relation to this everyday phenomenon we call technology' (553).

Rotblat (2001) sees the scientist and engineer as having a major voice in the future. At one level that is the voice of the individual engineer who, for instance, chooses not to work for technology that he or she perceives has or could have a negative effect on society or the environment, or an engineer who chooses to work for some firms specifically to contribute concern for creative responsibility. The very act of mining, for instance, affects the social and physical environment directly. Scientists and engineers can work on attempts to creatively balance these effects, as described in the case of Anglo American (Chapter 7).

In this sense we move to the view of the engineer as not being simply a 'number cruncher', but a as a professional who actively contributes through practice to the development of global ethical meaning. Rotblat takes this further, suggesting that every engineer or scientist should be a philosopher as well as a technical expert. Philosophy demands careful reasoning and provides 'different ways of looking at the world', beyond the simple area of competence or expertise. Rotblat then suggests that the engineer might formally take an oath before entering the profession. The very term 'profession' refers to a commitment or an oath that commits one to a task. Rotblat (2001, 174) quotes from the oath some American Universities ask their students to take, analogous to the Hippocratic oath for medical practitioners (no longer taken):

> I promise to work for a better world, where science and technology are used in socially responsible ways. I will not use my education for any purpose intended to harm

> human beings or the environment. Throughout my career I will consider the ethical implications of my work before I take action. While demands upon me may be great, I sign this declaration because I recognise that individual responsibility is the first step on the path to peace.

It may be objected that this provides too elevated a view of the engineer, but in another sense it takes the engineer seriously both as an individual and as a member of a profession that will in small or large ways affect the social and physical environment. Martin and Schinzinger (1989) suggest a range of several models of the engineer's profession role in relation to this:

- Saviour. The engineer as the key player in the creation of a utopian society with the development of technology and material prosperity for all.
- Guardian. If engineers cannot create a utopian society then they can help to guard against the development of a society ruled by technology. This involves helping to work through the best interests of society, based upon engineering knowledge.
- Social enabler and catalyst. Here the engineer does not simply take orders but works with managers and with society to help stakeholders to see the options and the values involved. In this way the management of technology revolves around creative partnerships, developing meaning as they go along.

The International Network of Engineers and Scientists for Global Responsibility (INES) argues the importance of a network of practitioners to promote such roles and responsibilities, principally:

> [T]o encourage and facilitate international communication among engineers and scientists seeking to promote international peace and security, justice and sustainable development and working for a responsible use of science and technology.

This includes:

- To work for the reduction of military spending and for the transfer of resources to the satisfaction of basic needs.
- To promote environmentally sound technologies while taking long-term effects into account.
- To enhance the awareness of ethical principles among engineers and scientists and to support those who have been victimized for acting upon such principles (from the INES Founding Statement, Berlin 1991).

The related organization, Scientists for Global Responsibility (SGR, http://www.sgr.org.uk), also provides material in helping to choose an 'ethical career' in engineering, including the issues involved in working for the arms industry and similar potentially problematic areas.

The Royal Academy of Engineering (RAE, http://www.raeng.org.uk) has also stressed the importance of bringing ethics firmly into the engineering curriculum in higher education. The RAE's Teaching Engineering Ethics Group (TEEG) has developed a map for embedding ethics in the curriculum. It is sometimes objected that the idea of teaching ethics is itself wrong, because it involves imposing an ethical view on the students. This is to misunderstand the point of ethics teaching, which is to enable the student to develop ethical autonomy in relation to their professional community. This takes into account both the ethical identity of the profession and the important responsibility of the individual profession to develop values that he or she can articulate and justify and the responsibility to make ethical decisions in practice. Hence, Rest (1994), for example, suggests four components of moral behaviour that can be developed through student-centred teaching:

- Moral sensitivity. Awareness of the situation and the stakeholders needs and values.
- Moral judgement. The capacity of moral reasoning that enables the person to identify and justify core values.
- Moral motivation. The prioritizing of moral values.
- Moral character. The development of moral qualities alongside skills (as noted in Chapter 3).

In whatever way the engineer styles himself or herself, there can be little excuse in the twenty-first century for not being aware of the global, environmental, organizational and professional issues that challenge the accountability and responsibility of the engineer and the engineering profession.

Questions

How do you now see your identity as a professional engineer?

Sum up your personal and professional ethical values.

Investigate the value statements, mission statements and CRS statements of six firms that you would consider working for.

What are the core ethical values and practices of these organizations and how do they relate to your values?

Conclusions

It is hard to confine ethics to a narrow area. On the contrary, we have seen that ethics is involved and focused in the profession of engineering, in the management of business and in the social context in which both operate. At one time or another, the engineer or the company may have ethical concerns in any of these areas.

At the core of much of our review has been the idea of responsibility and the importance of working this out in context. One cynical view of ethics in the past has been 'doing what you can get away with', precisely avoiding responsibility for actions. Plato tells the story of the ring of Giges, a ring that makes the wearer invisible. If we are invisible, do we need to take responsibility to be ethical? The answer from corporations such as Enron was seen to be 'no'. They thought that their actions were invisible. However, with globalization, the increase in the number of NGOs, the increase in political and legal concern for CSR, there are less and less places to hide. That in itself provides motivation to be ethical. But as other case studies have shown, there is an increasing concern for profession and business to take positive responsibility for their role in society and their effects on the social and physical environment.

None of this argues that any human organization can achieve completely ethical behaviour. As Bauman (1998) suggests in his description of the modern organization, its size involves inevitable division of labour, and this in turn leads to division of responsibility. With division of responsibility comes the loss of an overview and with that the loss of shared responsibility. It is at that point that we can feel ourselves approaching ethical invisibility. Such a dynamic is endemic to any organization. It will tend to recur, unless there is regular reflection on the purpose and practice of the engineer, on a professional, business and organizational level.

None of this is to impose values on the engineer or business. As Tawney (1930) reminds, there are no value-free social organizations. All embody moral values. The important thing is to know what those values are, to be able to justify them, and thus to be able to live with them. Then perhaps we can begin to see what it means to 'call yourself an engineer'.

References

Bauman, Z. (1998). *Modernity and the Holocaust*. London: Polity.

Bruntland, G.H. (1987). *Our Common Future – The World Commission on Environment and Development*. Oxford: Oxford University Press.

Carter, R. (1985). A taxonomy of objectives for professional education. *Studies in Higher Education*, **10**(2), 135–49.

Cooper, J. (2002). *Great Britons*. London: National Portrait Gallery Publications.

Cullen, F. (1987). *Corporate Crime under Attack: The Ford Pinto Case and Beyond*. London: Anderson.

Davies, P. (2001). Managing technology: some ethical considerations for professional engineers. In *Technology and Ethics: A European Quest for Responsible Engineering* (P. Goujon and B. Heriad Dubreuil, eds) pp. 543–54. Leuven: Peeters.

Ellul, J. (1964). *The Technological Society*. New York: Vintage.

Florman, S. (1976). *The Existential Pleasures of Engineering*. New York: St. Martins.

Illich, I. (1977). *Disabling Professions*. London: Marion Boyars.

Jonas, H. (1984). *The Imperative of Responsibility: In Search of an Ethics for the Technological Age*. Chicago: University of Chicago Press.

Martin, M.W. and Schinzinger, R. (1989). *Ethics in Engineering*. New York: McGraw Hill.

Pescatore, C. (1999). *Long-Term Management of Radioactive Waste, Ethics and the Environment*. OECD.

Porritt, J. (2006). *Is Nuclear Power the Answer?* Sustainable Development Commission.

Postman, N. (1993). *Technopoly: The Surrender of Culture to Technology*. New York: Vintage.

Rest, J. (1994). Theory and research. In *Moral Development in the Professions* (J. Rest and D. Narvaez, eds) pp. 1–26, Hillsdale, New Jersey: Lawrence Erlbaum.

Robinson, S. (2001). *Agape, Moral Meaning and Pastoral Counselling*. Cardiff: Aureus.

Robinson, S., Kendrick, K. and Brown, A. (2003). *Spirituality and the Practice of Healthcare*. Basingstoke: Palgrave.

Rotblat, J. (2001). Interview and Nobel Peace Prize acceptance speech. In *Technology and Ethics: A European Quest for Responsible Engineering* (P. Goujon and B. Heriad Dubreuil, eds) pp. 169– 82, Leuven: Peeters.

SDC Position Paper (2006). *The Role of Nuclear Power in a Low Carbon Economy*. Sustainable Development Commission.

Tawney, R.H. (1930). *Equality*. London: Allen and Unwin.

Wenk, E. (1986). *Tradeoffs: Imperatives of Choice in a High Tech World*. Baltimore: John Hopkins Press.

Index